TRANSLATION AND INFORMATION TECHNOLOGY

Translation and Information Technology

Edited by
Chan Sin-wai

The Chinese University Press

Translation and Information Technology
  Edited by Chan Sin-wai

© **The Chinese University of Hong Kong** 2002

ISBN 962–996–077–X

THE CHINESE UNIVERSITY PRESS
The Chinese University of Hong Kong
SHA TIN, N.T., HONG KONG
Fax: +852 2603 6692
      +852 2603 7355
E-mail: cup@cuhk.edu.hk
Web-site: www.chineseupress.com

Printed in Hong Kong

Contents

Part 3: Critique and Training

Translation and Information Technology: Machine and Machine-aided Translation in the New Century

Chan Sin-wai
Department of Translation
The Chinese University of Hong Kong

In recent years, translation studies has firmly positioned itself in the midst of the technological revolution. With communciation becoming faster and more efficient, translation also has to be done and delivered rapidly, accurately and in formats that fit the various modalities now available. These modalities pose new challenges for translation, and they exert an enormous impact on the development of the discipline on the practical, theoretical and educational levels.

In fact, the importance of machine translation and machine-aided translation in the present age of information technology cannot be over-emphasized. One of the primary functions of machine translation, as we all know, is to serve as a translation aid to remove linguistic barriers in order to cope with an ever-increasing volume of materials that appear on the Internet in languages that we do not know. The Internet, among other things, has actually revolutionized our way of life. The world has become a global village in which a multitude of languages have been used. Machine translation has made global communication possible. This is aptly pointed out by Colin Haynes (1998:3): "For the first time in history, we now have the capability to communicate with each other across the boundaries of time, space and language."

On a more microscopic level, machine translation is also important in regional development. With China's entry into the World Trade Organisation and to preserve Hong Kong's role as the world's leading financial centre, the demand for competent translators in this part of the world will be enormous. The shortage of human translators will be minimized by the supply of reliable domain-specific machine translation systems. This can be illustrated by a comparison between the amount of translation that can be produced by a human translator with that of a machine translation system. It is estimated that a human translator can, on average, translate 2,000 words per day (eight hours), but the speed of an average translation system is around 3,000 words per second. In eight hours, a system can translate as many as 86,400,000 words, which is 43,200 times faster than a human translator.

On the other hand, the quality and marketability of machine translation have been improved considerably. Machine translation software is becoming more and more viable, affordable, and user-friendly. The production of low-cost and easy-to-use software, coupled with the growing need to translate online information, means that machine translation systems will be widely used and web translation will be a leading trend in the future. In fact, machine translation software is more than an academic pursuit; it is a profitable business. According to Muriel Vasconcellos and L. Chris Miller (1996:1), "today, there are more than 500 vendors of machine translation software for the personal computer around the world, and among them they put out well over 1,000 products. One of the vendors, Globalink, sells its extensive line of software in at least 6,000 stores in North America alone, and at present Europe is its fastest-growing market." Obviously, machine translation is no longer a matter that is only of interest to academics, such as university linguists, computer engineers, and translationists, but an area from which commercial interests can be generated.

Despite this, machine translation systems in the market are far from satisfactory. One of the major reasons that accounts for the failure of machine translation is the over-dependence on the application of linguistic concepts to the construction of a machine translation system. The linguistic theories that have been used so far have not produced very effective and reliable systems. We certainly need to explore new areas and new methodologies to design new systems that really work. This basically has a lot to do with the distinction between what is known as a "language system" and "language in use." What the theorists have focused on are formalistic analyses of the grammatical and linguistic features of the languages concerned, to the neglect of empirical variations in actual usage within specific contexts. This accounts for the success of "practical systems" based on domain-specific grammars, and why a linguistics-oriented system with a complicated flowchart filled with technical jargon has not yielded results which are translationally acceptable.

It is true that a fully automatic high-quality machine translation system for general applications for any specific language pair still eludes us. For professional translators, machine-aided translation in the form of an interactive translation workstation will be extremely useful and is expected to become very popular in the near future. There is in fact no rivalry between machine translation and human translation because they complement rather than compete with each other. The human translator can help to improve the quality of machine output, and machine translation can speed up the process of human translation. Certain routine documents can be handled by the machine, while the human translator can spend more time on the production of quality translation of literary, legal and diplomatic documents.

Another point worth noting is that the design of a machine translation system has so far been in the hands of computer scientists and linguists. The views of translators, who are the professionals as well as the users, have not been taken into consideration in the construction of machine translation software. It is about time that human translators play a more active role in machine translation. And it can be expected that the role of translators in the design of a machine translation system will become increasingly important. A translational approach to machine translation will be a new development in the field and it can also be expected that a "translation-oriented" system with substantial input from practising translators in the areas of translation process, translation methodology and translation generation, will soon be in place.

The articles in this volume deal with many topics associated with machine and machine-aided translation, such as computational linguistics, bilingual corpus construction and the classification of terminology. The potential of computers in translation education is also explored and evaluated. Naturally, the Internet and its possibilities and challenges feature prominently here, as does the role of the human translator in a time of change.

Chan Sin-wai discusses *TransRecipe*, a machine translation system designed for translating Chinese recipes into English. This software applies various translation methods in the parsing stage to produce a high-quality result that eliminates the need for post-editing. Such software, he argues, can serve as a pilot system for the mechanical translation of procedural texts and practical writings such as instruction manuals, financial documents and scientific research.

Since computational linguistics form the basis for a successful machine translation programme, a range of papers deal with this topic. Liu Qun introduces a Chinese-English machine translation system based on micro-engine architecture. Chang Baobao and his fellow researchers discuss a joint project between Peking University, Tsinghua University and the Chinese Academy of Sciences, to develop a practical Chinese-English machine translation system designed as a multi-engine structure. A key component of this project is the setting up of a Chinese-English bilingual corpus. The authors discuss the design of the corpus, the principle on which the bilingual texts are captured, the composition of the corpus, its markup system, and the application of the corpus in a translation memory engine.

The paper given by Li Tangqiu and others presents a method of syntactic and semantic structure disambiguation for the Chinese language. The authors discuss the method used and give a brief introduction to its semantic knowledge resource — the *Hownet Dictionary*. Kit Chunyu, Pan Haihua and Jonathan J. Webster review the history of Example-based Machine Translation (EBMT) and discuss its main ideas. Major issues in EBMT have been discussed, including example acquisition, example matching and target sentence generation. The authors evaluate previous work on the subject and demonstrate the architecture of an EBMT system.

Jonathan J. Webster, Sin King Kui and Hu Qinan report on the application of semantic web technology in a study being carried out at City University of Hong Kong. They use the EBMT approach for the automation of natural language translation, with emphasis on legal document translation in Hong Kong.

Globalisation, technological innovation and the Internet have led to problems relating how terminology is translated from one language to another. Aman Chiu and Björn Jernudd deal with how terminologists, translators and IT specialists handle the Chinese translations of IT terminology during discourse. There has been a great proliferation of technical computing terms and acronyms due to the rapid development of Internet-related industries. Charlotte To and Björn Jernudd argue that without standardisation of these newly-created terms, whether in English or their translation equivalents, terminological problems are likely to result. The authors report on the terminological problems encountered by

Internet language professionals in Hong Kong through observations of real business meetings and focus group discussions.

The application of information technology to translation education is very apparent, and was raised by a number of conference participants. Beverly Adab discusses the use of electronic corpora within a tertiary setting and demonstrates how a right mix of theory and practice will equip students with the relevant analytical skills to adapt to the many different translation situations that they are likely to encounter in their future careers. Carrie Chau Kam Hung and Irene Ip Kwok Chun investigate the advantages and disadvantages of using computers in the learning of translation skills by conducting a study of junior and senior translation students from the Division of Language Studies at City University of Hong Kong.

The final series of articles look at the new technologies with the end-user in mind — the translator, the student-translator, and the translation client. The increasing interest in machine and machine-aided translation in the last two decades have led to a proliferation of English-Chinese software, all claiming to have the capability of producing high-quality translations. But how trustworthy are these claims? Li Defeng sheds some light on this question by means of a comprehension test carried out on texts produced by some English-Chinese translation software.

What is the challenge of machine translation to the traditional status of the translator? Paris Lau Chi-chuen discusses what happens to the notion of faithfulness, intelligibility and elegance when human subjectivity is overtaken by an automated machine. Evangeline S. P. Almberg goes on to examine work done by some of the state-of-the-art software for Chinese-English translation to see how much room is left for human hands to do.

On the basis of the articles in this collection, the major challenges for translation studies are how information technology impacts and alters previous thinking on translation theory, how it can be used effectively in the classroom and professional setting, and how to achieve better, more accurate results. The need for further research is obvious, and it is crucial that there be more collaboration between tertiary institutions, software companies and translation practitioners. The speed of change is breathtaking, and it is to translation's credit that that it has been at the forefront of this revolution.

In preparing this book for publication, I have been helped by the prompt delivery of papers by conference participants and the excellent input of Miss Jennifer Eagleton, my colleague at the Department of Translation, who spent many hours putting the manuscript in order for easy processing by The Chinese University Press.

References

Haynes, Colin (1998). *Breaking Down the Language Barriers*. London: Aslib.
Vasconcellos, Muriel and L. Chris Miller (1996). "Recent Trends in Machine Translation," in *Translating and the Computer 18: Papers from the Aslib Conference Held on 14 and 15 November 1996*. London: Aslib, pp. 1–5.

PART 1

Methodology and Application

The Making of TransRecipe: A Translational Approach to the Machine Translation of Chinese Cookbooks

Chan Sin-wai
Department of Translation
The Chinese University of Hong Kong

Introduction

TransRecipe is a fully automatic translation system for translating Chinese cookbooks into English based on the transfer model which combines the corpus-based, example-based, pattern-based, and rule-based approaches. With a lexical database of around 20,000 entries relating to Chinese food and drink, a general dictionary of around 2,000 frequently used expressions, a database of 200 examples, and a total of 700 global parsing rules, the system can automatically translate a recipe from Chinese into English within seconds without any pre-processing or post-editing.

This system sets out to achieve several goals in the field of machine translation which have not yet been realized by other Chinese-English translation systems on the market. Firstly, it will be the first fully automatic translation system specially designed for the rendition of Chinese recipes into English. Secondly, it applies a large number of translation methods in the parsing stage in order to produce a high-quality output that is translationally acceptable. Thirdly, it serves as a pilot system to illustrate the mechanical translation of procedural texts, which can have wider applications to other types of texts, such as instruction manuals, financial documents and scientific writings. This would be possible with the addition of a machine-tractable dictionary such as the author's electronic online dictionary entitled *A Translator's Chinese-English Dictionary*, which has 8,028 single-character entries, 75,559 multiple-character entries, and 213,457 English equivalents.

Apart from explaining issues relating to the making of *TransRecipe*, this paper intends to highlight the importance of using translation methods in the construction of a machine translation system to produce a good translation. This particular orientation is labelled as a "translational approach" to distinguish it from approaches based largely on linguistic and computational concepts. It is undeniable that the various machine systems for translating between Chinese and English, many of which are still far from satisfactory, have been designed mainly by those in the fields of computer science and linguistics and rarely by translation experts. As a result, most of the outputs are at best linguistically equivalent texts hardly intelligible to the reader. Also, as we all know, machine translation works at the syntactical level, and this means that methods frequently used in translating sentences in translation practice to produce effective translations have not been put to good use in machine translation. Methods of breaking down long sentences into shorter ones, for example, would certainly make it much easier for a machine translation system to produce a more accurate target text.

It was with the above two observations that I started three years ago to build a fully automatic translation system that renders Chinese recipes into English. The system will help the English-reading public in different parts of the world to practise the Chinese culinary art, which is a very important part of Chinese culture, and it will also show that the experience of a human translator can make a great deal of difference to the quality of machine translation.

Recipes: Grammatical Analyses

One of the most important steps in setting up a translation system is to analyse the source text linguistically, or more specifically, grammatically, since the sentence is the linguistic unit for a machine system. What is grammar? According to *Webster's Third New International Dictionary of the English Language* (Gove, 1971:986), grammar is defined as:

A branch of linguistic study that deals with the classes of words, their inflections or other means of indicating relation to each other, and their functions and relations in the sentence as employed according to established usage and that is sometimes extended to include related matter such as phonology, prosody, language history, orthography, orthoepy, etymology, or semantics.

According to one of the most influential works on English grammar, "grammar" is to "include both syntax and that aspect of morphology (the internal structure of words) that deals with inflections (or accidence)." (Quirk *et al.*, 1985:12) A lively description of grammar is given by Peter Newmark (1988:125) when he relates grammar to translation:

> Grammar is the skeleton of a text … Grammar gives you the general and main facts about a text: statements, questions, requests, purpose, reason, condition, time, place, doubt, feeling, certainty. Grammar indicates who does what to whom, why, where, when, how.

Translating, we must admit, is more than grammatical transfer. But it should also be emphasised that grammar can be used to break down the source language sentence for meaning acquisition and to verify the semantic correctness and syntactic structure of the target text. This explains why traditional grammar is used in this machine translation system. Let us discuss traditional grammer by different parts of speech (except for "interjection" which is never used in recipes).

(a) Adjectives

An adjective is a word that is used to modify a noun, expressing a characteristic quality or attribute. English adjectives can be both attributive and predicative, premodified by intensifiers and can take comparatives and superlatives. In recipes, one of the most commonly used adjectives is *shao xu* 少許, the translation of which depends on its collocates. The following equivalents have been used when translating *shao xu*.

麻油少許	*a dash of* sesame oil
荔枝醋少許	*a little* lychee vinegar
芫荽少許	*a pick of* parsley
胡椒粉少許	*a pinch of* pepper
鷹粟粉少許	*some* cornflour

Another point to note is that very often, adjectives in recipes are not translated according to the dictionary meanings. *Liang* 靚, for example, is usually translated as "beautiful." In Chinese recipes, it is translated as "fresh," such as translating *liang zhu ru* 靚豬肉 as "fresh pork."

(b) Adverbs

An adverb is a word or phrase whose main function is to modify a verb, an adjective, a sentence, or another adverb. There are several types of adverbs: adverb of degree, of direction, of frequency, of manner, of place, of recurrence, of sequence, and of time. Among them, the following two are relevant to recipes.

Adverb of Manner

This type of adverb, such as *carefully, easily, quickly, slowly*, refers to the manner something is or has been done. In the *TransRecipe* system, some manner adverbs in recipes are usually not translated, such as the word *tong* 同, meaning "together."

鹹豬肉、豬腿肉同飛水。
Scald salted pork and pork from hind leg in boiling water.

Adverb of Time

An adverb of time, also known as a temporal adverb, refers to the time something was or will be done. Adverbs such as *now, hourly, yesterday* are adverbs of time. Recipes inevitably have such kinds of adverbs as an indication of duration has to be included in cookbooks.

加入調味料炆十分鐘，加入冬菇再炆五分鐘。
Put in seasonings and braise for 10 minutes. Add black mushrooms and braise for another 5 minutes.

(c) Articles

There are three types of articles: definite article (*the*), indefinite article (*a/ an*), and zero article. The treatment of articles is in fact one of the most difficult issues for a human translator working especially from Chinese into English. A machine translation system will find it difficult to change an indefinite article to definite article after a noun has been introduced or identified in the previous sentence.

I saw a girl. The girl was wearing a red dress.

Fortunately for translators, only the definite article is used most frequently in recipes. It is the addition of the definite article that worries the rule-writer of the machine translation system.

(d) Conjunctions

Conjunctions are words which are used to join clauses together, and to show the relationship between the ideas in the clauses. The functions of conjunctions in both English and Chinese are more or less the same. Nevertheless, conjunctions in English are used more frequently than in Chinese. This is because the clauses in a Chinese composite sentence are usually connected by parataxis, whereas those in an English complex or compound sentence are connected by hypotaxis. The main difference between parataxis and hypotaxis lies in the fact that connectives are much less imperatively needed in a Chinese composite sentence than in an English complex or compound sentence.

The paratactic and hypotactic differences between conjunctions in Chinese and those in English should also be considered together with the use of the "set-off" punctuation mark, known as *dun hao* 頓號, in a Chinese text. A number of rules have been written to turn the Chinese paratactic sentence into a hypotactic structure.

將牛肉、豬肉、羊肉洗淨
wash clean the beef, pork and mutton

將牛肉、豬肉洗淨切片
sauté beef and pork and shred

冬菰、蝦米浸透切粒
soak black mushrooms and shelled shrimps thoroughly and cut into dices.

(e) Nouns

Words that name persons, places, things or ideas are nouns. Particular to recipes are nouns describing shapes. Recipes are full of nouns describing different shapes into which ingredients are cut.

剞菱形花紋
slash surface with criss-cross pattern

切成方形
cut into squares

切成十字形花紋
cut into criss-cross patterns

The following example serves to illustrate the above point:

麵粉篩勻放入大碗中，慢慢加入滾水，迅速拌勻成柔軟粉糰，以少許麵粉爽手，將粉糰揦成長條形，再分切小圓粒，碾薄成圓形粉皮，放入肉丸做成小籠包形狀。

Sieve and place flour in a large bowl. Gradually add boiling water. Stir quickly to make a soft dough. Knead into rectangular roll. Dice and press into thin round sheets. Place on meatballs. Make into the shape of Shanghai pork dumplings.

(f) Prepositions

A preposition is a function word placed before nouns, pronouns and gerunds to indicate a spatial or grammatical relationship. The following examples show that prepositions have to be handled contextually.

將蝦仁泡油   Soak shrimps in hot oil.

用滾水料滾水   Scald in boiling water with scalding ingredients.

(g) Pronouns

In Chinese, as in English, distinctions are made among *he* 他, *she* 她, *it* 它/牠 and *He* 祂 and agreement between nouns and pronouns is generally considered as one of the most difficult issues to handle since reference cannot always be found syntactically. Fortunately, recipes do not or need not contain pronouns which make the work of translation much easier.

雞洗淨，抹乾水斬件，加醃料醃十分鐘，泡油。
Wash the chicken clean, wipe dry, cut into pieces,
marinate for 10 minutes and scald in oil.

(h) Verbs

In translating verbs, a number of issues, such as tense, aspect, voice, have to be considered. Of these, tense is most important. As we all know, tense is the grammatical expression of time, and is divided into the present and the past. There is no formal tense to express future action, which is determined by modality rather than the projected time of the action. It should be noted that firstly, the concept of time and the grammatical tense belong to two different categories. Time exists independent of tense, which varies from language to language according to the linguistic expression of temperal relations. Secondly, tense forms in English are shown by variations in the morphological form of the verb, e.g. *I walk* and *I walked*, while the word forms of Chinese do not change, tense being expressed by adverbs of time 時間副詞, such as *xian zai* 現在 (now), *jiang lai* 將來 (in future), and *ceng jing* 曾經 (once) and auxiliaries 助動詞 such as *liao* 了, *zhe* 着, and *guo* 過.

In translating recipes, only the present tense is used. Since they are written in directive format, agreement between noun and verb is avoided and all verbs are in their basic forms.

拖水後瀝乾水分待用　　scald, drain and set aside for later use.

Recipes: Linguistic Analyses

Recipes are not much different from some directives, basic operating instructions and manuals, in which the terminology is relatively limited and the grammatical structure comparatively simple. Linguistically, recipes belong to the domain of procedural discourse. The main function of procedural discourse is to prescribe or tell the reader of the text the process of doing something. A typical text of a Chinese recipe and its translation (Yam, 1997:82) is given below to illustrate the linguistic characteristics of procedural discourse.

〈生爆鱔片〉

做法：
(1) 鱔魚去骨洗淨，用乾布抹淨，切成塊。
(2) 將鱔魚放入碗中，用二湯匙粟粉撈勻，放入熱油中炸透，撈起瀝乾油，盛上碟。
(3) 用約一湯匙油，爆蒜肉，倒下調味料，煮滾後，淋上鱔魚面，再放上芫荽，即可供食。

Deep-fried Eel Slices

Method:
(1) Remove bones of eel and wash thoroughly. Wipe dry and slice.
(2) Put eel in casserole and mix with 2 tbsp cornflour. Fry in hot oil, drain and serve.
(3) Sauté garlic with 1 tbsp oil. Stir in seasonings and bring to boil. Pour sauce over eels and garnish with parsley. Serve hot.

It can be seen from the above that recipes are directives, which are a type of imperative sentence. What are the characteristics of a directive sentence? According to *A Comprehensive Grammar of the English Language* (Quirk *et al.*, 1985:827), there are two special features. Firstly, it generally has no subject. Secondly, it usually has a main verb in the based form. Also, there are seven types of directives:

V:	Jump.
VO:	Open the door.
VC:	Be reasonable.
VA:	Get inside.
VOO:	Tell me the truth.
VOC:	Consider yourself lucky.
VOA:	Put the flowers on the table.

What is more, "the imperative verb lacks tense distinction and does not allow modal auxiliaries. The progressive form is rare, and the perfective even rarer."

Taken together, all these features make them pliable to machine translation. Firstly, directive clauses are short and this makes it easier to write syntactical rules for text generation with high accuracy. Secondly, all sentences are in simple present tense, which means the difficulty of having to work out the right tense for Chinese sentences is removed. Thirdly, as no subject is involved, the obstacle of noun and pronoun agreement in Chinese-English translation no longer exists. For this type of text, machine translation can be of sufficient quality to replace human translation. The only other factor that needs to be considered is that Cantonese cuisine has a large number of Cantonese expressions which should be dealt with computationally.

TransRecipe: **The Construction of Its Algorithm**

The first stage in the construction of this cookbook translation system is to produce its algorithm by simulating the working process of a human translator. *Visual Basic* programming language has been used to build up the algorithm which combines the corpus-based, example-based, pattern-based and rule-based approaches. For the generation of a target text, the input source text goes through four stages of processing: (1) The first stage is "example-based" processing where a database of examples in the source text language will be aligned with input sentences through an intralingual matching engine, and then translations of the aligned sentences will be produced by an interlingual matching engine. (2) The second stage is "corpus-based" processing where bilingual matching of individual words and phrases takes place between the source sentences and the machine-tractable general and specialised dictionaries. (3) The third stage is "pattern-based" processing where sentences with similar patterns will be matched bilingually. (4) The last stage is "rule-based" processing where rules for text generation will produce the translation of the source text. The following is the flowchart for *TransRecipe*:

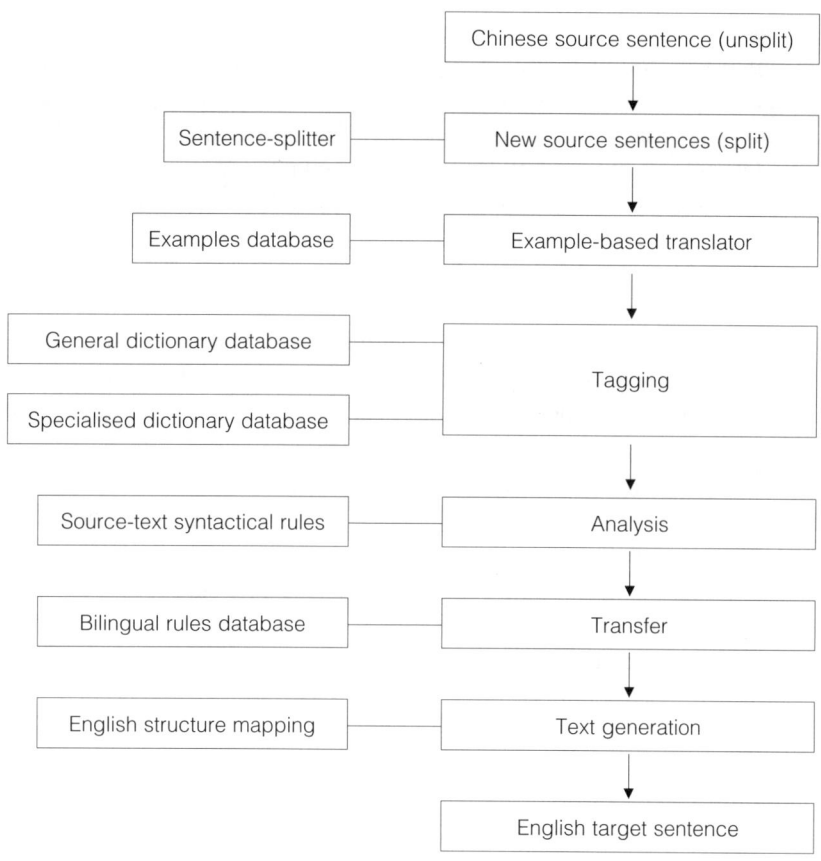

Flowchart of *TransRecipe*

Stage 1: Example-based Machine Translation (EBMT)

This is the most linguistic of the corpus-based approaches in machine translation. Example translations are also used as the basis of new translations. The idea was first suggested by Makoto Nagao in 1984 in his article published under the title "A Framework of a Mechanical Translation between Japanese and English by Analogy Principle." (Nagao, 1984) He applied this method in his translation system and described it as "translating through analogy." It was based on the idea that advances in computer technology have made it possible to gain access to the huge corpora of previously translated analogous examples to allow matching of bilingual expressions. It should be noted, however, that example-based machine translation cannot be wholly used for text generation as it is virtually impossible to translate every sentence by this approach. It is mainly a companion module to improve the overall quality of machine translation by serving as an alternative to the knowledge-based approach, and as an aid to the traditional rule-based methods. In other words, it should be used in conjunction with other empirical approaches to machine translation. A typical example of example-based machine translation is the *Pangloss Mark III* machine translation system.

With the use of examples, rules are not required. If the input sentence is aligned with a matching sentence, correct translation will be produced. If there is no matching, then correction is needed and the new translation will serve as a new example in the bilingual corpus. This approach, needless to say, is dependent on the storage of a large number of translation examples in the database to enable the matching process. It is generally held that EBMT will include a sentence-aligned corpus, then there is the intra-language matching between a stored source language chunk with an input chunk, the inter-language matching between the source language chunk with the target language chunk, and finally there is the chunk combination.

In my construction of the Chinese-English *TransRecipe* system, the example-based approach has been used at the first stage. At this stage, a database of examples in the source text language has been aligned with input sentences through an intralingual matching engine, and then translations of the aligned sentences will be produced by an interlingual matching engine. Stored in this database is a bilingual corpus of example translations so that the source phrases or sentences will be translated by the best-match algorithm built up in the system. The example-based approach is best suited for frequently used expressions which cannot be easily dealt with by rules. For example, "*qin cai si, gan sun si ge shao xu*" 芹菜絲、甘筍絲各少許 can be translated as "some shredded spring onion and a pick of flower-shaped carrot" and "*ma you, hu jiao fen ge shao xu*" 麻油、胡椒粉各少許 can be rendered into "a dash of sesame oil and a pinch of pepper." This approach saves the trouble of having to write a large number of rules to handle the collocational issues relating to the matching of different meanings of *shao xu* 少許 with the different cooking ingredients used.

胡椒粉、麻油各少許	a pinch of pepper and sesame oil
胡椒粉、麻油少許	a pinch of pepper and sesame oil
胡椒粉、酒各少許	a pinch of pepper and some wine
胡椒粉、生粉各少許	a pinch of pepper and cornflour
胡椒粉、熟油各少許	a pinch of pepper and cooked oil
胡椒粉、糖各少許	a dash each of pepper and sugar
薑片、陳皮各少許	some ginger slices and dried tangerine peel
薑片、蔥段各少許	some ginger slices and spring onion sections
薑片、蔥段少許	some ginger slices and spring onion sections
薑汁、紹酒、糖、生粉各少許	a little ginger juice, Shaoxing wine, sugar and cornflour

Stage 2: Corpus-based Machine Translation

This is the use of a pre-existing large corpus of lexical items or translated texts to construct a machine translation system. It is therefore closely related to the building of the lexical databases of the machine translation system. For *TransRecipe*, two lexical databases have been built: a "General Dictionary" and a "Specialised Dictionary."

(a) The General Dictionary

Frequently used terms of different parts of speech are stored in the General Dictionary arranged according to the Hanyu Pinyin alphabetical order. It is expected that over 3,000 lexical items will have to be included in order to make the translation system work. Collected in this database are commonly-used expressions in Chinese cookbooks, such as *shao liang de* 少量的 "a small amount of" (adjective), *jun yun di* 均勻地 "evenly" (adverb), *ji* 及 "and" (conjunction), *wu fen zhong* 5 分鐘 "5 minutes" (noun), *yi qie wei er* 一切為二 "halve" (verb), and *guan* 罐 "can" (measure-word).

(b) The Specialised Dictionary

As recipes are normally made up of dish-names, ingredients and methods of preparation and cooking, the bilingual specialised dictionary database will be formed by terms in these three areas. The names of national and regional dishes form a major part of the bilingual database and terms related to the preparation of dishes have been collected and translated. At present, about 10,000 items of Chinese food and drink and their translations, classified into 112 categories and given grammatical tags, have been stored in the database. Names and translations of dishes and soups, meat, beancurd, fish, shellfish, vegetables, and poultry, dimsum, ingredients, condiments, sauces, desserts, rice, noodles, congee, kitchen equipment, tableware, and restaurants, have been included in the database. Problems that have been encountered in the compilation of this database include the treatment of countable and uncountable nouns, the indication of measure-words for all the common and proper nouns, the handling of variant characters with the same semantic meaning (e.g *shui fen* as 水份 and 水分), the use of simplified and standard characters (*sa shang* as 洒上 and 灑上) in the source text, and the translation of different expressions bearing the same meaning (*zhu gun* 煮滾, *zhu zhi gun* 煮至滾 and *zhu zhi gun qi* 煮至滾起 "bring to the boil").

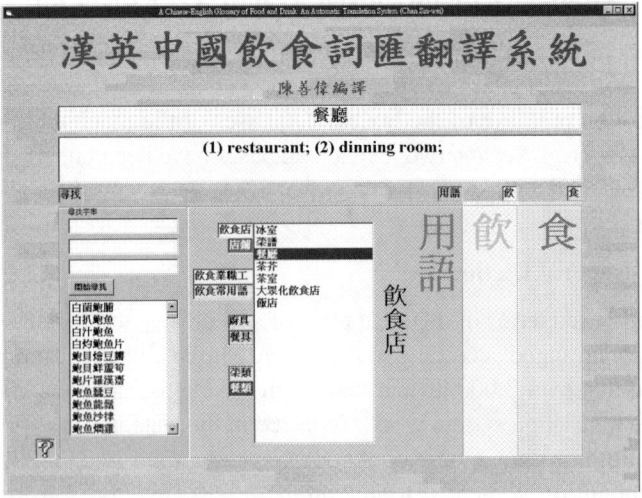

A Chinese-English Glossary of Food and Drink:
An Automatic Translation System

To make this Specialised Dictionary readily accessible to the public, it has now been made into a CD-ROM in which a number of search engines have been designed to facilitate the search of a specific entry.

Stage 3: Pattern-based Machine Translation

The use of sentence patterns to deal with sentences which vary only in a small number of variables, such as the direct object or objects of the verb, has become increasingly popular. With the addition of the "partial matching method," pattern-based translation will greatly enhance the quality of machine output.

浸軟 soak until tender
泡軟 soak until tender

粉絲浸軟。
soak mungbean vermicelli *until tender*.

花生用油炸成金黃色。
deep-fry peanuts in oil until golden.

火腿用溫水沖洗過。
rinse ham in warm water.

牛肉用清水浸十分鐘。
soak beef in water for ten minutes.

Stage 4: Rule-based Machine Translation

Rule-based Machine Translation (RBMT) is a relatively traditional method which depends greatly on the difficult and time-consuming work of preparing and maintaining a large number of rules and a huge amount of lexical information in the form of dictionaries, both general and specialised.

In translating between Chinese and English, this is used to convert the syntactic structures of Chinese sentences into the structures of the equivalent English sentences by the use of some reordering rules, which rearrange the words or characters in the source text in the order of the target text. The rule-based approach is therefore serving the important role of bridging the structures of the two languages.

Two types of rules have been written for the parser of the system: those treating the measure-words in the source text and those dealing with grammatical structures.

(a) Measure-word Rules

Rules governing the translation of measure-words have been written to deal with the items mentioned in the "Ingredients" and "Seasoning" parts of the recipe. According to the views of the linguist, a measure-word, or *liang ci* 量詞, is a type of classifier in Chinese that is placed between a numeral and a noun to indicate the unit of the object, a linguistic feature that is characteristic of the Sino-Tibetan language family. This type of measure-word, which refers mainly to the ways things are counted, is known collectively as "nominal measure-words" (*ming liang ci* 名量詞) or "object measure-words" (*wu liang ci* 物量詞), such as *yi ben shu* 一本書 (a book), *san tou niu* 三頭牛 (three cows). The other general type of measure-words is "verbal measure-words" (*dong liang ci* 動量詞), which refers to the way actions are counted, such as *shuo yi pian* 説一遍 (say it again), *zuo yi tang* 走一趟 (go there once).

In translating recipes, only the nominal measure-words need to be treated. Nominal measure-words can generally be divided into (1) "weights and measures" (*du liang liang ci* 度量量詞), such as *chi* 尺 (foot) and *jin* 斤 (catty); (2) "general unit measures" (*yi ban dan wei liang ci* 一般單位量詞), referring to the way things are normally counted, such as *yi ge ren* 一個人 (a person), *yi zhang zhi* 一張紙 (a sheet of paper); and (3) "container measures" (*rong qi liang ci* 容器量詞), referring to the way things have been contained or held, e.g. *yi wan fan* 一碗飯, *yi tong shui* 一桶水. Another way to classify nominal measure words is to put them into two categories based on their grammatical structures: (1) simple measure-words (*dan ci liang ci* 單詞量詞), such as *zhang* 張 and *chi* 尺; and (2) compound measure-words (*fu ci liang ci* 複詞量詞), such as *ren ci* 人次 (person-time) and *jia ci* 架次 (sortie).

(i) General Measure-words

A large number of measure-words fall into this category. As each Chinese common noun carries with it a measure-word, the number of measure-words to be treated in the database is expected to be huge. In reality, this is not the case as most weight measures have been used for individual nouns, which greatly minimize the need to translate a huge number of measure-words for each common noun.

(ii) Container Measure-words

There are two types of container measure-words: single container measure-words, such as *bei* 杯 (cup), and compound container measure-words, such as *cha chi* 茶匙 (tsp). These measure-words differ from others in their countability. All containers are countable common nouns. Some may be expressed in a form which does not need to indicate its singularity or plurality, such as *tsp* or *tbsp*. Others may need entries in the lexical database to indicate their singular and plural forms.

There are various forms of expression of measure-words:

上湯六杯　　　6 cups of stock
上湯或水六杯　6 cups of stock or water

(iii) Weight Measure-words

Weight measure-words are mostly used in the "Ingredient" and "Seasoning" sections of a recipe. Weights in recipes are usually given in terms of the Chinese traditional system 中國舊制 of *qian* 錢, *liang* 兩 and *jin* 斤 and the British system of ounces and pounds. These systems need to be converted into the metric system.

Several methods have been used to translate Chinese measure-words.

(1) Method of Amplification

This method is applied to fill grammatical or textual gaps when translating from one language into another. As measure-words, which are strictly speaking different from classifiers, do not exist in English, Chinese-English translation involves the insertion or omission of proper measure-words before nouns.

薑四厚片　　4 thick slices ginger
薑一厚片　　1 thick slice ginger

紅蘿蔔二條　2 carrots
紅蘿蔔一條　1 carrot

(2) Method of Direct Equivalence

The method of direct equivalence is used for "container measures" 容器量詞 as *bei* 杯 is translated as "cup" and *cha chi* 茶匙 as "teaspoon."

蛋汁三湯匙
3 tbsp whisked egg

蛋汁一湯匙
1 tbsp whisked egg

(3) Method of Conversion

The method of conversion is used when turning *jin* 斤 and *liang* 兩, units in the Chinese traditional system, into either the metric system in terms of grams (g) or British system in terms of ounces (oz).

牛肉一斤
600 g beef

Chinese		Metric
一兩	1 tael (*liang*)	= 40 g
一斤	1 catty (*jin*)	= 600 g

British		Metric
一安士	1 oz	= 30 g
一磅	1 lb	= 450 g

(4) Method of Collocational Translation

More problematic is the translation of the general measure-words which may or may not be translated, depending on the collocates, and which may be translated into a number of equivalents, depending on the nouns with which measure-words are associated.

伊麵二個	2 cakes E-fu noodles
伊麵一個	1 cake E-fu noodles
蒜二粒	2 cloves garlic
蒜一粒	1 clove garlic

(5) Method of Multiple Translations

Measure-words have a wide range of collocations. Rules have to be written to contextualize the translation.

一棵芫荽	a sprig of parsley
一棵西蘭花	a stalk of cauliflower

(b) Grammatical Rules

Grammatical rules, other than the rearrangement of word order in translating the source text into English, involve the use of the following translation methods.

(1) Method of Addition

The addition of the conjunction "and"

The "set-off" punctuation mark (、) is usually converted into a comma when translated into English. But used between two items, it has to be turned into a conjunction, which has to be added to the rules.

再將牛肉、豬肉加入 add beef *and* pork

The addition of the definite article "the"

將牛肉洗淨   wash *the* beef

The addition of conjunction and article

將牛肉整條洗淨瀝乾 wash clean *and* dry *the* beef

(2) Method of Omission

Omission is a translation method which leaves some words in the original untranslated to achieve grammatical accuracy and idiomaticity in the translation. Omission is used in *TransRecipe* for the following purposes:

(i) To achieve syntactic completeness after breaking down the sentences according to preset rules.

當飯熟後   Cook the rice.
(當 and 後 have not been translated)

(ii) To achieve brevity when the sentence is procedurally comprehensible.

再將牛肉放入鑊
Put the beef into the pan.
(再 and 將 have been omitted)

然後放入蜜桃煮約十分鐘
Put in honey peach and boil for 10 minutes.
(然後 has not been translated)

(3) Method of Fronting

As with all imperative sentences, verbs in their base form are always placed at the beginning of the sentence. Grammatical rules for recipes have therefore been written to front the first verb in the Chinese source sentence in the English translation. For Chinese sentences beginning with a verb, no fronting is necessary, as in:

放入牛肉、豬肉及羊肉煮至滾起。
Put in beef, pork and mutton and bring to the boil.

But when the first verb in the source sentence is not in the initial position, then fronting is necessary and rules should be written to put the translation into a proper imperative sentence in English. The following are Chinese imperative sentences that begin with different parts of speech and their first verbs have been fronted in the translations:

將青豆及磨菰飛水 (*adv.*)
scald green beans and button mushrooms

與蝦仁同用少許生粉撈勻 (*conj.*)
stir well shelled shrimps with some cornflour

蝦連殼洗淨 (*noun*)
wash the shrimps with shells

用小火將材料焗至熟 (*prep.*)
bake ingredients over low heat until cooked

其餘原隻蒸熟 (*pron.*)
steam the rest until cooked.

Stage 5: Testing of the System

When the lexical and syntactical databases are in place, the system can be put into use. Around 500 recipes have so far been tested using the system, and it is estimated that 1,000 recipes have to be tested before *TransRecipe* can be put on the market. This testing process is extremely important for the evaluation of the system in the areas of accuracy, clarity, speed and reliability, and also for improvements to be made on the effectiveness of the parser and the enhancement and optimization of the various databases.

Demonstration of the *TransRecipe* System

Source Text

〈燴牛仔骨〉

材料：
牛仔骨四至五條，洋蔥一個，番茄二個，蒜茸、乾葱蔥茸各一茶匙
醃料：
生抽3/4湯匙，酒半湯匙，胡椒粉少許

調味料：
茄汁三湯匙，生抽1¹/₂茶匙，鹽1/3茶匙，糖二茶匙，水1¹/₂杯
做法：
(1) 牛仔骨斬成大段，放入醃料拌勻。
(2) 洋蔥去外衣，切成大塊，略泡油待用。
(3) 洗淨番茄，切大塊，去籽後再洗淨，瀝去水分，待用。
(4) 燒熱二湯匙油，爆香乾蔥茸、蒜茸，加入番茄塊略炒，放入洋蔥、調味料及牛仔骨煮至滾起。
(5) 將上項滾起的材料用中火燴煮至材料腍及汁濃，灑入胡椒粉少許，即可供食。

TransRecipe Output

Stewed Short Ribs

Ingredients:
4–5 beef short ribs. 1 onion. 2 tomatoes. 1 tsp each of mashed garlic and minced shallots
Seasonings:
3/4 tbsp light soy. 1/2 tbsp wine. A dash of pepper
Seasonings:
3 tbsp tomato sauce. 11/2 tsp light soy. 1/3 tsp salt. 2 tsp sugar. 11/2 cups water
Method:
(1) Chop beef short ribs into sections. Marinate with seasonings.
(2) Peel onions. Cut into large pieces. Fry briefly in hot oil and set aside.
(3) Wash clean tomatoes. Cut into large pieces. Seed and wash clean. Drain. Set aside.
(4) Heat 2 tbsp of oil. Sauté minced shallot and mashed garlic. Add tomato and stir-fry for a while. Place, onions, seasonings and beef short ribs and bring to the boil.
(5) Place boiled ingredients over medium heat and stew until cooked and sauce thickens. Sprinkle with a dash of pepper. Serve.

We must, of course, bear in mind that while each system has its own strengths, few of them produce satisfactory results when applied to real world texts. Logically, a hybrid system would fare better than a single-approach system. And it is quite likely that domain-specific systems will prevail in the near future as general application or all-purpose systems have yet to demonstrate their workability in a variety of contexts.

Conclusion

It goes without saying that as we are now in the age of information technology, translation will inevitably be widely used to overcome cross-lingual problems in data extraction and consumption. Web translation, as a result, has become increasingly popular. The main obstacle that cannot be easily solved is that no programming language, no matter how powerful it is, is capable of fully and sufficiently processing the myriad multifaceted information on the Internet written in the major natural languages in the world. This means that while it is important to achieve globalisation with the use of web translation, considerable allowance must be given to its inadequacies and inperfections due to the inability of an artificial language to match with a natural language. But for domain-specific texts, the matching of languages is possible as its limited vocabularies and syntactical variations make them susceptive to machine processing. The relatively smooth running of *TransRecipe* demonstrates that with the application of translation methods commonly used in translation practice, a new "translational approach" is an alternative to the traditional linguistic approach. With this new approach, and with the use of the machine-tractable dictionary, *A Translator's Chinese-English Dictionary*, we are ready to march on into other domains. And as we do so, a general application system will eventually emerge. Then, and only then, can machine translation remove the linguistic and cultural barriers that have so far hindered global communication.

References

Gove, Philip Babcock, ed. (1971). *Webster's Third New International Dictionary of the English Language*. Springfield, Mass.: G. and C. Merriam Company.

Nagao, Makoto (1984). "A Framework of a Mechanical Translation between Japanese and English by Analogy," in Alick Elithorn and Ranan Barnerji, eds., *Artificial and Human Intelligence*. Amsterdam: North-Holland Publishing Company, pp. 73–80.

Newmark, Peter (1988). *A Textbook of Translation*. Hertfordshire: Prentice-Hall.

Quirk, Randolph, Sidney Greenbaum, Geoffrey Leech and Jan Svartvik (1985). *A Comprehensive Grammar of the English Language*. New York: Longman Group Ltd.

Yam, Lisa 方任利莎 (1997). 《方太食譜之魚蝦蟹》 (*Lisa Yam's Cook Book: Seafood*). Hong Kong: Ming Pao Press Ltd.

A Chinese-English Machine Translation System Based on Micro-engine Architecture

Liu Qun
Institute of Computing Technology, Chinese Academy of Sciences
Institute of Computational Linguistics, Peking University

Introduction

Despite the use of a number of technologies in the design of Machine Translation (MT) systems, none of them has produced an optimal output on free text. A multi-engine MT approach has therefore been proposed to integrate several MT engines in one system. (Frederking and Nirenburg, 1994) Such an approach, as shown in the following diagram (Figure 1), has been successfully used by a number of MT systems (Frederking *et al.*, 1994:73–80; Frederking, Rudnicky and Hogan, 1997:61–65; Nirenburg, 1996:96–105; Rayner and Carter, 1997:107–10). Experiments have shown the results of using multi-engine MT system are indeed better than any of the single MT engines in the system (Hogan and Frederking, 1998).

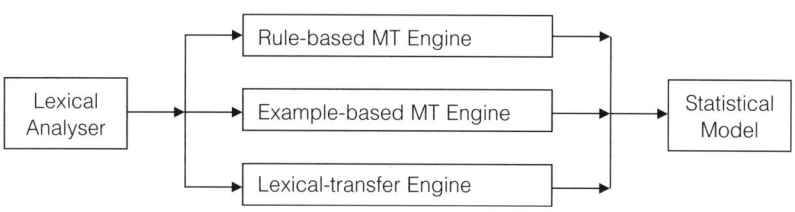

Figure 1. Structure of Multi-engine MT

In such a system, each engine tries to translate the source sentence separately, gives a series of translations of the phrases in the source sentence, and then puts the resulting output segments into a shared chart-like data structure. All the partial translations can then be given an internal quality score. A chart-walk algorithm is used to find the best combination of the partial translation.

In multi-engine architecture, the engines work independently. That means, an engine cannot make use of the results of other engines. For example, an Example-based MT (EBMT) engine can translate a Chinese sentence "我喜歡看電影", because there is a sentence "我喜歡看電視劇" in the corpus. But for the sentence "我喜歡看成龍演的這部電影", the EBMT engine cannot give the result, because there is no sample in the corpus can match the phrase "成龍演的這部電影". It is possible that a rule-based MT (RBMT) engine can translate this phrase correctly. But in the multi-engine system, the EBMT engine cannot use the result given by the RBMT engine.

Here we give a micro-engine approach to machine translation and introduce a Chinese-English machine translation system using such an approach. Similar to the multi-engine approach, it can synthesize the results of different MT engines. What is more, engines in the micro-engine system can interact with each other.

The Micro-engine Architecture

A micro-engine MT system consists of several micro-engines and an engine manager. All the micro-engines share a chart data structure. The engine manager also maintains an active constituent list. An active constituent is a constituent recognized by a micro-engine but has not been used to generate new constituents. The engine manager selects the best active constituent from the active constituent list and sends it to all the micro-engines. The micro-engines recognize new constituents using this active constituent and the existing inactive constituents (the edges in the chart). The engine manager will add these new constituents to the active constituents list and the previous selected active constituent will be moved from the active constituent list to the chart. This process repeats itself until a constituent covering the whole input sentence is recognized.

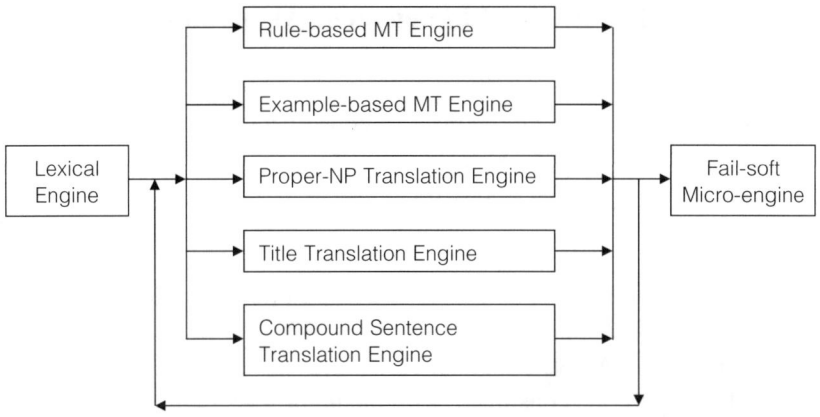

Figure 2. Structure of Micro-engine MT Architecture

The Micro-engine

A micro-engine is a machine translation engine. Unlike a traditional MT engine, a micro-engine does not try to translate the whole input sentence. A micro-engine is specialised. It only tries to find a specific type of constituent in the input sentence and translates these constituents. All the engines work cooperatively to translate the whole input sentence. That means, an engine can make use of the results generated by other micro-engines.

Normally, a micro-engine should perform two functions:

(1) *Recognize*

The micro-engine accepts an active constituent, combines it with the existing inactive constituents, and generates a list of new constituents.

(2) *Translate*

The engine should translate the constituent it recognizes. It may call the *Translate* function of the micro-engines that recognize the sub-constituents.

The Engine Management Algorithm

Data:

Chart — containing all the inactive constituents
ActiveList — the active constituent list
EngineList — the list of micro-engines

Algorithm:

Use the lexical engine to recognize all the words in the source sentence
Add these words into ActiveList
Repeat while ActiveList is not empty
    TheEdge = the constituent with the highest score in ActiveList
    If TheEdge covers the whole input sentence
        Call the *Translate* function of TheEdge
        Return the result translation text
    EndIf
    For EachEngine in EngineList
        Call the *Recognize* function of EachEngine using TheEdge as
           input
        Add all the output constituents to ActiveList
    EndFor
    Remove TheEdge from ActiveList
    Add TheEdge to Chart
    Sort ActiveList according to certain criterion
EndRepeat
If no constituent covering the whole sentence is recognized
        Use the Fail-soft Engine to find a best combination of existing
           constituents
    Translate the constituents in the combination
    Return the result translation text
EndIf
EndAlgorithm

Lexical Engine and Fail-soft Engine

Normally, if the input active constituent is empty, the micro-engine's *Recognize* function will not do anything. But there are two exceptions. These two micro-engines are called the Lexical Engine and the Fail-soft Engine.

The Lexical Engine is the engine that carries out the lexical analysis. That means, to lookup the dictionary, to segment the Chinese sentence to words, and to recognize Chinese personal names and place names. Its *Recognize* function works only when the input active constituent and the chart are both empty.

The Fail-soft Engine is used when there is no constituent covering the whole input sentence recognized. It selects the best combination of the existing constituents in the chart and generates the translation based on them. Its *Recognize* function works only when the input constituent is empty and the chart is not empty.

Other Micro-engines

In addition to the Lexical Engine and the Fail-soft Engine, we use five other micro-engines in our Chinese-English machine translation system.

One of the micro-engines is a Rule-based Engine. This engine is constructed from a traditional rule-based Chinese-English machine translation system. (Liu and Yu, 1998:514–17) There are about 300 syntax rules in this engine. It uses a chart-parsing algorithm to parse the sentence.

Another micro-engine is an Example-based Engine. We collected a bilingual corpus of about 200,000 words. Most texts in the corpus consist of news or editorials from the Xinhua News Agency or Government White Books.

The third micro-engine is a Proper-NP Translation Engine. This engine can recognize proper noun phrases from the Chinese sentence and translate them into English. The proper noun phrases include person name phrases, place name phrases, organisation name phrases, time phrases, number phrases, money phrases, and so on.

The fourth micro-engine is a Title Translation Engine. The titles of Chinese articles usually have a special syntax structure, such as "機器翻譯的預處理研究", "試論網絡黑客的行為方式" and "魯迅傳". This micro-engine can recognise this kind of titles and translate them properly.

The fifth micro-engine is a Compound Sentence Translation Engine. This engine can find the logical relations between the simple sentences in a compound sentence according to the conjunction words, and translate the sentence properly.

An Example

Here is an example to show how the micro-engine MT system works. For a Chinese sentence:

> 演員帕特里克・斯威茲在他最近的一部電影中扮演了一個感人的保鏢角色。(Actor Patrick Swayze played a touching bouncer in one of his recent movies.)

The Lexical Engine will look up the dictionary and cut the sentence into words:

> 演員/n 帕/g 特/g 里/f 克/v ・/w 斯/g 威/g 茲/g 在/p 他/r 最近/a 的/u 一/m 部/q 電影/n 中/f 扮演/v 了/u 一/m 個/q 感人/a 的/u 保鏢/n 角色/n 。/w

The labels following each words, such as "n," "g," "w," are the part-of-speech tags of the words.

The Proper-NP Translation Engine will recognize the "帕特里克・斯威茲" as a transliteration of a foreign name:

> 帕特里克・斯威茲/n (Patrick Swayze)

The Rule-Based MT Engine will recognize and translate the constituents as below:

> Np 演員帕特里克・斯威茲 (Actor Patrick Swayze)
> Pp 在他最近的一部電影 vp (in his recent a movie)
> Vp 扮演了一個感人的保鏢角色 np (played a role of a touching bodyguard)
> S 演員帕特里克・斯威茲在他最近的一部電影中扮演了一個感人的保鏢角色。(Actor Patrick Swayze played a role of a touching bodyguard in his recent a movie.)

The resultant translation is not so grammatical in English.

While the Example-based MT Engine can translate some phrases in another way:

> Np 在他最近的一部電影中 vp (in one of his recent movies)
> Vp 扮演了一個感人的保鏢角色 np (played a touching bodyguard)

These partial translations are better, because the Example-based MT Engine can translate phrases by comparing them with similar examples in the corpus, rather than translate them according to the manually written rules.

Finally the Rule-based MT Engine will synthesize the intermediate result to produce an accepted translation:

S 演員帕特里克・斯威茲在他最近的一部電影中扮演了一個感人的保鏢角色。(Actor Patrick Swayze played a touching bodyguard in one of his recent movies.)

In this example, we can see that different micro-engines work cooperatively and the translation is better than what may be generated by any of the single engines.

Conclusion and Future Work

Both the micro-engine MT system and the multi-engine system can employ different MT technologies within a single system. But there are still differences between them.

The granularity of engines in a micro-engine system is finer than that of engines in a multi-engine system. In a multi-engine system, each engine should be a complete MT system. It tries to translate the whole input sentence. But in a micro-engine system, each engine has its specialty. A micro-engine does not need to try to translate the whole sentence, it just needs to translate the "familiar" part of the sentence and ignore the rest of the sentence.

A micro-engine system is a close-coupling system. In such a system, all the engines work cooperatively. The relation between the micro-engines is cooperative, rather than competitive and an engine can take the intermediate results of other engines as its input. In contrast, a multi-engine system is a loose-coupling system. Engines in such a system work separately.

The micro-engine architecture of machine translation has a strength that other architecture cannot match. It is easy to develop. It is true that the engines can work cooperatively in a micro-engine system, the programming interface between micro-engines is, however, rather simple. The micro-engines need not handle the complicated communications between them. The system also has good scalability, so adding new micro-engines to a micro-engine system will not cause modification of the old system. And because the micro-engines need not translate the whole sentence, developers can focus their attention on a specific problem such as improving accuracy. Lastly, the engine management algorithm can be easily modified to become a parallel algorithm, which will take the advantage of a rapid parallel computer that has several CPUs. It is our intention to develop more micro-engines in the future to improve our MT systems.

References

Frederking, Robert and Sergei Nirenburg (1994). "Three Heads Are Better than One," in *Proceedings of the Fourth Conference on Applied Natural Language Processing* (ANLP-94). Stuttgart, Germany, pp. 95–100.

Frederking, Robert, *et al.* (1994). "Integrating Translations from Multiple Sources with the *Pangloss Mark III* Machine Translation System," in *Proceedings of the First Conference for Machine Translation in the Americas* (AMTA). Columbia, Maryland, October, pp. 73–80.

Frederking, Robert, Alexander Rudnicky and Christopher Hogan (1997). "Interactive Speech Translation in the DIPLOMAT Project," in Steven Krauwer *et al.*, eds., *Spoken Language Translation: Proceedings of a Workshop*. Madrid, Spain: Association of Computational Linguistics and European Network in Language and Speech, July, pp. 61–65.

Hogan, Christopher and Robert Frederking (1998). "An Evaluation of Multi-engine MT Architecture," in David Farwell, Laurie Gerber and Eduard Hory, eds., *Machine Translation and the Information Soup*. New York: Springer, pp. 113–23.

Liu, Qun and Yu Shiwen (1998). "TransEasy: A Chinese-English Machine Translation System Based on Hybrid Approach," in David Farwell, Laurie Gerber and Eduard Hory, eds., *Machine Translation and the Information Soup*. New York: Springer, pp. 514–17.

Nirenburg, Sergei, *et al.* (1996). "Two Principles and Six Techniques for Rapid MT Development," in *Proceedings of the Second Conference of the Association for Machine Translation in the Americas* (AMTA). Montreal/Quebec, Canada, October, pp. 96–105.

Rayner, Manny and David Carter (1997). "Hybrid Processing in the Spoken Language Translator," in *Proceedings of ICASSP-97*. Munich, Germany, pp. 107–10.

Bilingual Corpus Construction and Its Management for Chinese-English Machine Translation

Chang Baobao, Zhang Huarui, Yu Shiwen
Institute of Computational Linguistics, Peking University

Kang Shiyong
Department of Chinese Language and Literature,
Yan Tai Normal College

Introduction

In recent years, mono- or multilingual corpora are viewed as key resources in language information processing and language engineering projects. A large number of new approaches to language-related applications and research based on large-scale corpora have been proposed. Aligned bilingual sentence pairs, for example, can be directly used as a translation memory to improve the quality of the machine translation, and useful data or knowledge can also be extracted from a bilingual corpus based on a statistical model, such as the acquisition of bilingual translation patterns. On the other hand, bilingual corpora can also be very valuable to bilingual lexicographers and linguists.

There are three interrelated fields concerning the study of multi- or bilingual corpora: (1) techniques and methods to process or annotate collections of bilingual texts. Many proposals to tag, parse and align bilingual corpus have been published and more and more programmes or tools for such purposes have appeared (Gale *et al.*, 1993); (2) models based on multi- or bilingual corpora to specific applications in which multiple languages are involved. For example, Brown and Nagao (Brown *et al.*, 1990; Nagao, 1984) used bilingual corpora in different ways to develop machine translation system and Klavans and Tzoukermann (1990) are interested in utilising bilingual corpora in lexicography; (3) general issues in designing, compiling and encoding of corpora. TEI (Text Encoding Initiative) and CES (Corpus Encoding Standard) based on SGML (Standard Generalised Markup Language) are being developed to markup text structures. Most researchers in China have focused their attention on the first two fields. (Liu, Zhou and Huang, 1995) Systematic and well-encoded bilingual corpora, especially with Chinese as the source language, is not yet available.

The Institute of Computational Linguistics at Peking University, the National Key Laboratory for Intelligence Technology at Tsinghua University and the Institute of Computational Technology of the Chinese Academy of Sciences are jointly developing a practical Chinese-English machine translation system which has been funded by the Chinese Government since January 2000. For the purpose of combining all the benefits of different translation methods, the system is designed as a multi-engine system. In this paradigm, the traditional rule-based engine, corpus-based engine and other translation engines will coexist and interact in the final system. As a key resource, a Chinese-English bilingual corpus with about one million Chinese characters and 0.6 million words of corresponding English texts is being set up. In this paper, the design, collection, markup and annotation of such a corpus are described in detail.

The Design of the Corpus

When compiling a new bilingual corpus, even if it is small, it is sensible for compilers to design it carefully. Compilers must make decisions on the corpus type, size and composition. A valuable corpus is not an arbitrary collection of arbitrary texts. Careful planning will ensure that the large amount of work involved in compilation is worthwhile.

As to our task of constructing a bilingual corpus, we think it is very important to always keep intended usage of the corpus in mind. The corpus will serve a Chinese-English machine translation system, which will mainly deal with newspaper news texts found in the Internet for assimilation purposes. We make it clear that our corpus is a specialised corpus instead of a general one. A high coverage of all text types will make no sense, in contrast to a corpus which is dedicated to linguistic analysis and research. The content of the corpus, the text categories, the structure of the corpus, the sources from which texts are collected, and the time when the texts are produced, should all be suitable for translating newspaper news. It would be ideal if the corpus could be a sample of the total population of news texts so that it is statistically significant. However, it is quite difficult to construct a corpus which can meet all the theoretical requirements mentioned above. The first problem we faced is that there are not enough bilingual newspaper texts to collect and translation of Chinese newspaper is very expensive. We therefore had to make a trade-off between theoretical criteria and the availability of news texts. Finally, we collected texts according to the following principles:

(1) Ideally the best text type was news reportage (which we preferred), but some other materials similar to news texts with good translation could also be included in the corpus. In addition to news reportage, we also collected texts of press conferences, essays and policy papers as well as their translation produced by the Chinese government.

(2) All bilingual texts should have Chinese as the source language if possible, for serving a Chinese-to-English machine translation system. However, some existing easy-to-collect English sentences with professional Chinese translation were also introduced into the corpus. For example, about 25,000 English-Chinese sentence pairs derived from a machine translation evaluation project (Yu, Jiang and Zhu, 1991), where such sentence pairs serve as test sets, are included in the corpus.

(3) All texts should be collected as full texts, but there are also exceptions. We think full texts will be a useful resource to learn more on text structure in the future.

(4) All texts to be collected in the corpus should be recently published.

Guided by these principles, we have collected about one million Chinese character full texts and their English translations, mainly from the Internet, and about 35,000 sentence pairs. All full texts are one of four types — news reportage, transcripts of press conferences, policy papers and essays. The composition of the full text corpus is illustrated in Figure 1 (page 34):

The Markup of the Corpus

The best way to manage parallel texts is to develop special tool programmes, which will make the corpus easy to use. But that means that all texts from different sources must be encoded or marked up in the same

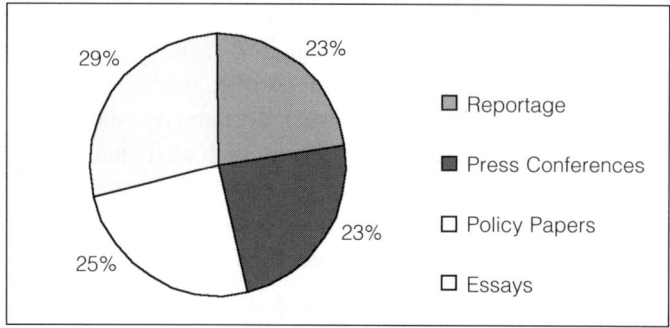

Figure 1. The Composition of the Full Text Corpus

way. A uniform representation will ensure that the corpus is easy to interchange across applications and software platforms. There are two famous corpus-encoding standard proposals available — the Corpus Encoding Standard (CES), still under way to its final version, and the Text Encoding Initiative (TEI), which have been used by some monolingual corpus such as the British National Corpus (BNC). Both markup standards are based on the Standard Generalised Markup Language (SGML). Because most texts in our corpus come from the Internet, most of them are originally encoded in Hyper Text Markup Language (HTML). We could also choose to develop a new encoding system. The best way to markup our texts could only be determined by careful consideration and comparison of these four selections.

Firstly, HTML form was abandoned, even those it seemed that the least work is needed to encode all texts. And it seems that downloaded text could be used directly without any modification. But this is incorrect. HTML is a widely used markup language used by websites and has many variations. Software enterprises, such as Microsoft and Netscape, have made different extensions to and inserted new elements into it. The syntax of HTML is not strict, therefore many web pages contain errors and express the same meaning in different forms. Most importantly, HTML is a presentation and content-mixed markup system. There are both content tags, such as <Hn>, and presentation tags, such as <font>. Generally speaking, authors of web pages do not use content elements so as to achieve a special display effect. For example, they prefer to use <center> and <font> to make the titles of texts more eye-catching instead of using tags, such as <Hn>.

Both CES and TEI are designed to encode a corpus. But the problem is that both of them are designed for general purposes. Both of them are difficult to grasp and use, even when we only select a minimum set of necessary elements. Some of what they consider to be necessary tags have less relevance to our needs. But some elements used to tag the structure of news reportage are not available. Furthermore, both CES and TEI are based on SGML, which is proven to be too complex to use and is not widely employed by those working in the area of information technology. To develop a fully functional SGML parser is no easy task.

In order to achieve a simple, but workable encoding solution suitable for our purposes, we chose to develop a new tagging system with reference mainly to CES. The new system does not try to cover all document types, but has the necessary tags for news reportage documents, and only with general support to other document types. We also based our system on the currently most popular markup language today — extensible Markup Language (XML), which is a small subset of SGML and supported by many important software companies.

According to our encoding system, the whole corpus is composed of sets of interrelated documents. The logic structure of the bilingual corpus is shown in Figure 2.

In this paradigm, a Chinese text and its English translation are represented by five cross-linked documents:

(1) *Document containing the Chinese text with basic information*
   In this document, the structure of the original Chinese text is

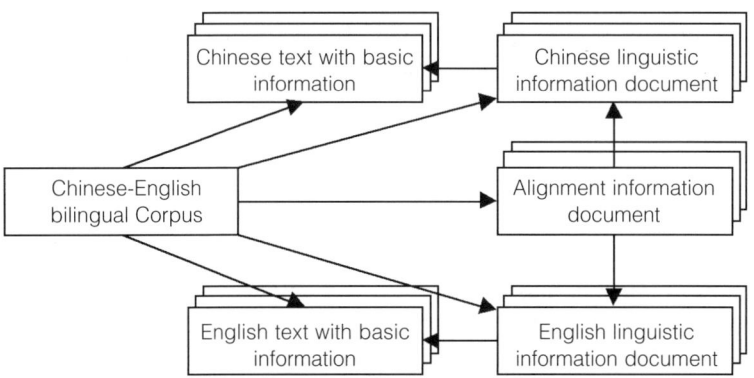

Figure 2. The Logic Structure of the Bilingual Corpus

tagged with predefined tags. The title, subtitle, author, reportage introduction, communication head (such as "新華社，9月10日北京") and other elements of news reportage or general documents are tagged. Named entities, such as personal names, organisational names, are also tagged in this document.

(2) *Document containing the English text with basic information*
This document is similar to the first document — documents containing the Chinese text with basic information. The only difference is that this document is for the English text.

(3) *Chinese linguistic information document*
In this document, linguistic information about words, phrases, sub-sentences and sentences of the Chinese text is recorded.

(4) *English linguistic information document*
In this document, linguistic information about words, phrase, sub-sentences and sentences of the English text is recorded.

(5) *Alignment information document*
Alignment information about the Chinese text and English text is recorded in this document.

There are Four Document Type Definitions (DTDs) defined for representing all documents mentioned above. One DTD file is for general description of the whole bilingual corpus, another DTD file is for alignment information document, documents containing the Chinese text with basic information and documents containing the English text with basic information share yet another DTD file, and the last DTD file is for Chinese linguistic information document and English linguistic information document.

With this markup system, simple annotation and deep annotation can be tagged in a uniform way. This also makes step by step annotation possible.

The Annotation and Alignment of the Corpus

Concerning corpus annotation, the first issue to consider is how annotations should be carried out on a corpus. That undoubtedly depends on how the final corpus will be used. Our corpus is supposed to be used in Chinese-English machine translation, and we hope part of the corpus can be used directly or indirectly as resource in the Chinese-English machine translation system. We also hope useful translation knowledge and statistical data could be extracted from the corpus, such as a bilingual lexicon and its translation pattern. The second issue is there must be an efficient way to annotate the corpus. Corpus annotation is very time-consuming, labour-intensive, expensive and sometimes error-prone.

At present, we are carrying out or planning to carry out the following kinds of annotations to the corpus:

(1) Chinese word segmentation and pos-tagging;
(2) English tokenisation and pos-tagging;
(3) Chinese and English proper noun identification is also planned (a small-scale experiment on Chinese organisational names has been carried out);
(4) alignment of the Chinese text and the English text at sentence level;
(5) alignment of Chinese proper nouns and their English translations is also planned;
(6) tagging the Chinese word with detail grammatical features with regard to the context. Such an annotation is based on the *Grammatical Knowledge Base of Contemporary Chinese*. (Yu and Zhu, 1996) In the knowledge base, words of a particular type have dozens of grammatical features and possible values. But with regard to context, only one or small subsets of the values of features make sense in authentic texts. Such annotation will make it possible to learn word translation rules. So far, we have conducted some experiments on such annotation.

The typical procedure for conducting the annotations is as follows: if there are available programme tools, first annotate the corpus automatically; then, verify the machine-tagged corpus by human professionals. Using Chinese word segmentation and tagging software, the Chinese text of the corpus is segmented and pos-tagged according to specification made by the Institute of Computational Linguistics, Peking University. With an English tokenisation programme we developed and modified a freely downloadable pos-tagger, the English text was tokenised and tagged according to *Penn Treebank* tagset. An improved existing alignment programme was used for the alignment of bilingual texts at the sentence level. We have also developed a rudimentary Chinese organisational name identification programme based on statistical knowledge, but the experiment result has not been good (with recall = 69.7%, accuracy = 40.9%). Improvements are required before it comes into use.

So far, 10% of the full text corpus has been tokenised, tagged and verified. Verification of the results of the automatic sentence alignment is under way. Figure 3 shows samples of the annotated corpus.

Figure 3: Samples of the Annotated Corpus

(A) *Fragment of Tagged Chinese Text*

[香港/ns 特別/a 行政區/n]ns 成立/v 以來/f ，/w 香港/ns 繼續/v 保持/v 著/u 亞太地區/j 重要/a 的/u 國際/n 金融/n 、/w 貿易/vn 、/w 資訊/n 、/w 航運/n 中心/n 和/c 世界/n 最/d 大/a 的/u 自由港/n 地位/n ；/w 世界/n 最/d 大/a 的/u 集裝箱/n 碼頭/n 一/w 香港/ns 葵涌/ns 貨櫃/n 碼頭/n ，/w 平均/a 每/r 天/q 有/v 112/m 艘/q 遠洋/b 巨輪/n 進出/v 。/w 亞洲/ns 最/d 大/a 的/u 香港/ns 新機場/n ，/w 已/d 投入/v 運營/v ；/w 截至/v 今年/t 第一/m 季度/n ，/w 香港/ns 外匯/n 儲備/vn 968億/m 美元/q ，/w 穩/ad 居/v 世界/n 第三/m 位/q ；/w 香港/ns 社會/n 人心/n 穩定/a 。/w 前/f 些/q 年/q 移居/v 海外/s 的/u 港人/n 大量/m 回流/v 。/w 今年/t 5月/t ，/w [香港/ns 特區/n]ns 第一/m 屆/q 立法會/j 選舉/v ，/w 148萬/m 居民/n 踴躍/ad 參與/v ，/w 投票率/n 高/a 達/v 53%/m ，/w 超過/v 港/j 英/j 時期/n 立法局/n 的/u 投票率/n ；/w 香港/ns 社會/n 秩序/n 良好/a 。/w 據/p 香港/ns 警方/n 統計/v ，/w 特區/n 成立/v 以來/f ，/w 香港/ns 社會/n 的/u 整體/n 犯罪率/n 是/v 24/m 年/q 來/f 最低/a 的/u 一/m 年/q ，/w 全年/n 刑事/b 案件/n 比/p 上年/t 下降/v 了/u 15%/m 。/w 去年底/t 一/m 項/q 社會/n 調查/vn 顯示/v ，/w 有/v 79%/m 的/u 人/n 感覺/v 在/p 香港/ns 生活/v 非常/d 安全/a 。/w

(B) *Fragment of Tagged English Text*

Since/IN the/DT founding/NN of/IN the/DT SAR/NNP ,/, Hong/NNP Kong/NNP has/VBZ maintained/VBN its/PRP$ status/NN as/IN an/DT important/JJ financial/JJ ,/, trade/NN ,/, information/NN and/CC shipping/NN center/NN in/IN the/DT Asia-Pacific/NNP Region/NNP ,/, and/CC the/DT largest/ RBS free/JJ trade/NN port/NN in/IN the/DT world/NN ./. Kwai/NNP Chung/NNP Pier/NNP ,/, the/DT world/NN 's/POS largest/JJS container/NN wharf/NN ,/, berths/VBZ on/IN the/DT average/JJ 112/CD ocean-going/JJ vessels/NNS every/DT day/NN ./. Hong/NNP Kong/NNP 's/POS new/JJ airport/NN ,/, the/DT largest/JJS of/IN its/PRP$ type/NN in/IN Asia/NNP ,/, has/VBZ started/VBN operation/NN ./. By/IN the/DT end/NN of/IN March/NN, /, Hong/NNP Kong/NNP 's/POS foreign/JJ exchange/NN reserves/NNS had/VBD totalled/VBN US/NNP $/SYM 96.8/CD billion/CD ,/, steadily/RB ranking/VBG third/JJ Cin/IN the/DT world/NN ./.

(C) *Fragment of Aligned Ttexts*

{{{{
[[(1:1)
(%s)第一，在人民幣不貶值的條件下，立足自己，發揮優勢，降低成本，增加出口的競爭能力。
%$
(%s)First, under the condition that the Renminbi (RMB) will not devaluate, we will rely on own efforts and advantages to lower cost and increase the competitiveness of our exports.
]]
[[(1:1)
(%s)降低成本主要是依靠自己，特別是依靠出口企業從過去的粗放經營向集約化經營轉變。
%$
(%s)In this regard, we will chiefly rely on transforming the management of export-oriented enterprises from an extensive to an intensive pattern.
]]
}}}}

Further Work and Application of the Corpus

A lot remains to be done on the corpus, such as a deeper study of annotation, which is not being currently planned given the lack of supporting tools and its complexity.

Ten percent of the corpus has been tokenised, tagged and aligned at sentence level, so the task of tokenising, tagging and aligning the other ninety percent of the corpus is underway.

There are many ways to use the corpus in a Chinese-English machine translation system. It is shown here how a Translation Memory (TM) engine that we are developing uses the corpus.

Bilingual sentence pairs are the key resource of the translation memory engine. The sentence pair can be either simply annotated or deeply annotated. At present, the TM engine is supposed to have two different levels of sentence annotation. At the first level, the Chinese sentence and English sentence are aligned, but the sentences are not tokenised and tagged. For example, alignment of 10% of the full text corpus produces 2,534 sentence pairs, and these sentence pairs are being indexed to form a simple translation memory in which there is no other information besides alignment information. With such a translation memory, the TM engine can only provide a search operation. If the input of the TM engine happens to be in the memory, the output translation will be produced directly.

At the second level, the Chinese sentence and English sentence are aligned, but the sentences are also tokenised and tagged, and proper nouns are also aligned. Such a collection of sentence pairs can also be indexed to form a translation memory. Because the translation memory contains much more information than simple translation memory, the TM engine will view the sentence pair as a sentence level translation pattern with the aligned proper nouns as pattern slots. For example, Sentence pair (1) will correspond to pattern (2):

(1) 應克林頓的邀請，中國總理朱鎔基將於4月6日至14日訪問美國。
    At the invitation of President Clinton, Chinese Premier Zhu Rongji will visit the United States between April 6–14.

(2) 應X1 X2的邀請，X3 X4 X5將於X6 X7至X8訪問X9。
    At the invitation of X2' X1', X3' X4' X5' will visit X9' between X6' X7'–X8'.

Such sentence level translation patterns give the TM engine the ability to substitute proper nouns. For example, the TM engine will perform the following translation by utilising the above pattern and bilingual lexicon:

INPUT: 應江澤民主席的邀請，馬來西亞總理馬哈蒂爾將於10月3日至6日訪問中國。

OUTPUT: At the invitation of President Jiang Zemin, Malaysian Premier Mahathir will visit China between October 3–6.

Acknowledgments

I would like to thank Prof. Duan Huiming, Dr. Chen Yuzhong, Dr. Wu Yunfang, Prof. Liu Qun and some other colleagues and students in the Institute of Computational Linguistics, Peking University for their contribution in verifying the machine-tagged corpus. Thanks also should be given to Dr. Wang Bin of the Institute of Computational Technology of the Chinese Academy of Sciences, who provided the initial programme for aligning the corpus at sentence level. Improving an existing tool rather than developing a new one saves much work. It should also be noted that such a work could not have been done by authors alone. Thanks should also be given to other colleagues who contributed to this paper but whose names have not been listed above.

References

Brown, P., *et al.* (1990). "A Statistical Approach to Machine Translation." *Computational Linguistics*, Vol. 16, No. 2, pp. 79–85.

Corpus Encoding Standard, http://www.cs.vassar.edu/CES/.

Gale, William, *et al.* (1993). "A Program for Aligning Sentence in Bilingual Corpora." *Computational Linguistics*, Vol. 19, No. 1, pp. 75–102.

Klavans, Judith L. and E. Tzoukermann (1990). "The BICORD system," in *Proceedings of the 15th International Conference on Computational Linguistics*, pp. 174–79.

Liu, Xin, Zhou Ming and Huang Changning (1995). "Experiment on an Alignment Algorithm for Chinese-English Parallel Texts Based on Sentence Lengths," in Chen Liwei, ed., *Progress and Application in Computational Linguistics*. Beijing: Tsinghua University Press, pp. 62–67.

Nagao, Makoto (1984). "A Framework of a Mechanical Translation between Japanese and English by Analogy Principle," in Alick Elithorn, *et al.,* eds., *Artificial and Human Intelligence*. Amsterdam: North-Holland Publishing Company, in cooperation with NATO Scientific Affairs Division.

TEI Guidelines for Electronic Text Encoding and Interchange, http://etext.virginia. edu.

Yu, Shiwen, Jiang Xin and Zhu Xuefeng (1991). "An Automatic System for Evaluating Machine Translation," in *Proceedings of MMT'91*, pp. 57–58.

Yu, Shiwen and Zhu Xuefeng (1996). "The Specification of *The Grammatical Knowledge Base of Contemporary Chinese*." *Journal of Chinese Information Processing*, No. 2, pp. 1–22.

A Hybrid Method for Syntactic and Semantic Structure Disambiguation of Chinese

Li Tangqiu, Yang Xiaofeng, Hong Qingyang and Li Shaozi
Department of Computer Science, Xiamen University

Introduction

In sentence parsing, a proper structural and semantic disambiguation is very important. But the problem has not been well tackled. It is well known that there are two major sources of ambiguity. One is lexical (polysemy), while the other is structural and semantic. Traditionally, two major methods are used in the disambiguation. One is based on constraint resolution. The other is based on semantic preference. The former rules out some possibilities using some rules of syntactic or semantic constraints. The latter picks out the most preferable one according to some syntactic or semantic preferences. In practice, these two methods are often combined. For example, a sentence may be parsed first according to some syntactic rules, and then one of the most favourable candidate structures is chosen as the final parse according to various factors.

Usually, it is unable to deal with structural ambiguity depending solely on the syntax. For example, *"wei xiu tu shu guan di kong tiao"* 維修圖書館的空調 and *"zhuang xiu tu shu guan di gong ren"* 裝修圖書館的工人 are similar in surface form. But their syntactic structures are quite different according to their possible meaning. One is (*wei xiu (tu shu guan di kong tiao*)) (維修 (圖書館的空調)) and the other could be (*(zhuang xiu tu shu guan) di gong ren*) ((裝修圖書館) 的工人). To solve this problem, we must appeal to semantic knowledge. That is *kong tiao* 空調 (air-conditioner) cannot act as the subject of *zhuang xiu* 裝修 (doing maintenance), and *gong ren* 工人 (worker) could not be the object of *zhuang xiu* 裝修 (decorate). A semantic network may be used to solve such a problem, if it can contain all the possible and impossible relationships between various concepts. But building such a semantic network with completeness and accuracy is not feasible at present.

In Natural Language Processing (NLP), a thesaurus is often taken as the source of semantic knowledge. But the purpose of an ordinary thesaurus is not for NLP, and its system structure and semantic category are not very suitable for the purpose of Chinese syntactic structure disambiguation.

The disambiguation method presented in this paper takes *Hownet* as its semantic knowledge resource. It compares the collocations of the notional words in the syntactic structures to be disambiguated with the collocations of the notional words in the examples in *Hownet*, and calculates their scores. The structure with the highest score will be taken as the final result. By using this parsing method, we can solve the problem of lexical and structural disambiguation.

Disambiguation Based on Knowledge

The method introduced here finds the most plausible interpretation of a phrase or a sentence in Chinese by evaluation of the similarity between the structure and the examples existed in the related word entries in a knowledge base, *Hownet*. We say it is a hybrid because it utilises the knowledge in a way that example-based systems utilise examples in a large-scale bilingual corpus, but the corpus is not built solely using statistical methods. Instead, it is carefully built by human experts and implemented with human expertise.

Broadly speaking, words in Chinese can be classified into two kinds: notional words and auxiliary words, according to the degree of the richness of sense. Notional words include nouns, verbs and adjectives. They have real meaning and there may exist many kinds of relationships between them. But most auxiliary words, on the other hand, can only express conjunctional relationship, tense, mood and so on. The semantic collocation relationship between notional words can play a very important role in the syntax structure disambiguation. Our evaluation of some candidates is based on the degree of tightness of match between notional words in the structure. In the lexicon, we provide representative collocation examples with strong discriminating capacity for each entry. These examples are phrases or simple sentences containing the word. We compare the context-related word-set of the word in the current structure with all the examples of the word in the lexicon, and find the best match. Then the best example is taken as the word's explanation. The degree of similarity shows how the word in the structure matches with other notional words in it, so it can be taken as the reference of the notional words. Because the discrepancy of different candidate parses of a structure, the same word has a different content-related word-set, and so will get different scores. We can calculate the best match according to the score of all the notional words of the sentence. Because we choose the entry with the best degree of similarity when comparing examples, we have also solved the most of the lexical disambiguation of the notional words at the same time.

To obtain the context-related word-set of a notional word in current structures, the parsing results should be represented in the form of a dependency tree. The root of the tree is the core word of the structure, and its predominated words are the children nodes of the root. The children trees are also dependency trees rooted on each children node. For example the input sentence *"wei xiu/tu shu guan/di/kong tiao"* "維修/圖書館/的/空調" can have different analyses, and can be represented in two dependency trees:

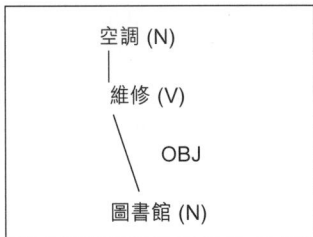

Figure 1. The Tree Structure of
(*(wei xiu tu shu guan) di kong tiao*)
((維修圖書館)的空調)

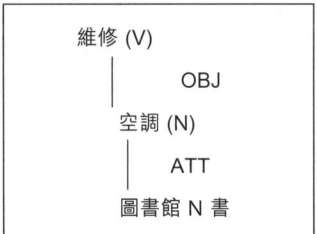

Figure 2. The Tree Structure of
(*wei xiu (tu shu guan di kong tiao*)
(維修(圖書館的空調))

Every node in the two trees represents a word and its part of speech. The arc between nodes shows the predominated relation between two related words.

The Semantic Knowledge Source

The Introduction of Hownet

Hownet is the great contribution of a senior researcher, Dong Zhendong, who spent nearly ten years building it. *Hownet* is a knowledge base, which defines the concept of words in Chinese and English, and shows the relationship between concepts and their properties. As its name implies, it is actually a large semantic network.

The semantic lexicon is the major part of the *Hownet* system, in which each semantic entry of a word forms a record. Each record includes four slot and value pairs connected by "=". On the left of "=" is the domain name and the right its value:

NO.   = serial number of words or phrases
[W_X = word or phrase itself
G_X   = part of speech of word or phrase
E_X   = examples of word or phrase]+
DEF   = concept definition

Here X in W_X、BG_X、E_X mentioned above is used to describe the different languages, with C standing for Chinese and E for English. The concept definition of each word is presented in DEF, the value of which is described as several semantic atoms and their semantic relation mark before them. A semantic atom is the most fundamental unit of what is called semantic primitives. *Hownet* extracts over 800 semantic atoms by observing and analysing more than 6,000 Chinese characters, and summarizes about 16 relations such as part, agent, theme, belonging, time, space, material, and so on. All these relationships are represented by prefix marks such as "*", "@", "$" symbols before semantic atoms.

We use a formalised language to define DEF:

DEF   = [Mark]Atom[,[Mark]Atom]*
Mark = * | @ | ? | ! | ~ | # | $ | % | ^ | &
ATOM = $atom_1$|$atom_2$|...|$atom_k$

All these semantic atoms and their relations are used to define concepts of all the words.

In designing *Hownet*, the E_X slot (examples of a word entry) are emphasized mainly on the discriminating capacity, not the explaining capacity. The examples in the slot are well chosen so that they can help word meaning disambiguation.

The following example is the definition of the word entry *da* 打 in the lexicon, whose meaning roughly corresponds to "play" in English when it is used in the phrase of *da qiu* 打球 in Chinese. Its grammar category is a verb:

NO.   = 017140
W_C = 打
G_C = V
E_C = ~網球，~牌，~秋千，~太極，球~得很棒
W_E = play
G_E = V
E_E =
DEF = exercise|, sport| 體育

The "~" sign in the E_C slot of the above example stands for the present word *da* 打. The examples in the E_C slot (~網球，~牌，~秋千，~太極，球~得很棒) are carefully chosen to discriminate this usage of *da* 打 from others.

Furthermore, *Hownet* also provides semantic definitions for each word entry (such as *exercise* | and *sport* | 體育 in the DEF slot in the above example). These semantic atoms, combined with taxonomy trees provided by *Hownet*, give the semantic definition and its relation with other concepts. Given two concepts, the taxonomy tree can be used to calculate the semantic distance between them. The following picture is the EVENT tree that presents relations of event semantic atoms. *Hownet* has EVENT, ENTITY, and a few other taxonomy trees.

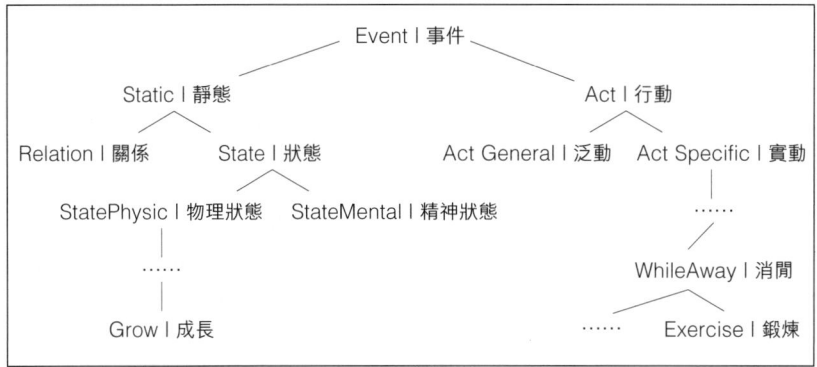

Figure 3. Event Taxonomy Tree

Semantic Calculation

The Dependency Tree of a Structure and the Context-window of a Word

First, assume that the input sentence S has been syncopated beforehand as S=(R1 R2 R3 … Rm), in which Ri (1_i_m) presents the *ith* phrases of the sentence. Assume also that the parent of the Ri node in the tree is Parent (i), and the set of children nodes is children (i), that the part of speech of Ri is Cat (I), and the semantic relation with its parent is Case (i). Then the node Ri can be represented as:

Node$_i$: (i Parent(i) Children(i) Ri Cat(i) Case(i)) (1_i_m)

This way we can transform a parse tree of a sentence into a new form of representation Tree={Node$_1$ Node$_2$ … Node$_m$}.

Secondly we define the determinative relations of notional words. Assuming A, B are notional words, if A modifies or dominates B in the sentence, then A is the determining word of B and B is the determined word of A. Note here that the determinative relation is different from the traditional dependency relation. In the traditional dependency relation if A modifies B, B is the dominating word and A is dominated word, hence A is attached to B. In other words, A is the child of B, while in our determinative relation definition, the determined word is modified and predominated object. For instance, for a sentence with subject-predicate-object structure, the verb is the determining word of the notional subject and object of the sentence; In ADJ+NP, the ADJ phrase is the determining word of the NP. If the determining and determined words have modifying relation, such as ADJ+NP, NP+NP, and the determining word acts as the attached composition of Attribute, Manner and so on, the determining word then is the child of the determined word in the dependency tree. If the sentence has a subject-predicate structure, such as NP+VP+NP, NP+ADJP, the determining word acts as the parent of the determined word.

The reason for us to distinguish the determining word and the determined one is that they have different effects in our algorithm. Determining words have greater inference on the match operation between words. In the algorithm, the scores of the determining words are calculated by the comparison of the example set of every word entry with its determined notional word set; while the scores of determined words will be calculated after the entry of the determining word is chosen. We usually choose the highest score as the evaluation reference.

Evaluation of a dependency tree is based on the evaluation of the notional words that can act as determining or determined words in the current structure evaluated.

Third, for a notional word which can act as a determining word, we define its context-related word-set, or context-window, as its all determined word set. We compare the context-window of a notional word with every example of each entry of the word, and choose the word entry with the highest degree of similarity as its semantic explanation. The highest score of the entry is taken as the score of the current word evaluated. For a notional word that only acts as the determined word in the current structure, we calculate its score based on the comparison result between the context-window and the example set of its determining word. We also define the score of the current structure as the total score of all these nodes included in it.

Given a syntax structure of a sentence, by using the corresponding dependency tree we can get the context-window of every word. The following is selection principles of the context-window of a notional word in the tree:

(a) If a notional word is attached to another word through the link labelled as MANNER, TENSE, TOOLS, ATTRIBUTE, choose the parent.
(b) If the children of a notional word are attached through a link labelled as SUBJECT, OBJECT, choose these children.
(c) If a word in a verb's window is the subject of the verb (this can be known by the syntax parsing), label it with subj.

According the above principle we can obtain the context-window of nodes in Figure 1 and Figure 2 as follows:

Table 1. The Context-window of the Nodes in Figures 1 and 2

Node	Window in Figure 1	Window in Figure 2
維修	空調	SUBJ，圖書館	空調
圖書館	NULL	空調
空調	NULL	NULL

Evaluation of Nodes

The evaluation of nodes involves the calculation of three degrees of similarity:

(1) semantic similarity between semantic atoms;
(2) semantic similarity between words;
(3) semantic similarity between the context-window of the word and the examples of each entry of the word.

These three calculations can be done using the three parts of knowledge of *Hownet* respectively, that is, taxonomy tree, DEF of a entry, and examples of each entry.

Calculation of the Degree of Similarity Between Semantic Atoms

> *Definition:* semantic distance between semantic atom a and b
> DISTANCE $-$ ATOM(a,b) = the shortest distance between a and b in taxonomy tree (1)
> *Definition:* the degree of similarity between semantic atoms
> SIM $-$ ATOM(a,b) = (1-DISTANCE-ATOM(a,b) / (height-of-taxonomy $-$ tree $\times$ 2)) $\times$ 100 (2)

Calculation of the Degree of Similarity Between Words

Definition: Assuming Relations(i) is the set of all the prefix relations of all the semantic atoms in the DEF of a word entry I, and Item-Relation-Atoms (i,K) is the set of all the atoms whose prefix relation is K, then the degree of similarity between entry I,J is calculated using following formula:

$$\text{SIM} - \text{WORD} - \text{ITEM}(I,J) = \sum_{k \in \text{Relations}(I)} \max_{\substack{a \in \text{Item–Relation–Atom}(I,k) \\ b \in \text{Item–Relation–Atom}(J,k)}} \text{SIM} - \text{ATOM}(a,b) \quad (3)$$

$$\text{SIM} - \text{WORD}(R1, R2, C) = \begin{cases} 0 & \text{if not } Cat\,(I2) = CAT\,(I1) = C \\ \max_{\substack{I1 \in Items\,(R1) \\ I2 \in Items\,(R2)}} \text{SIM} - \text{WORD} - \text{ITEM}(I1, I2) & \\ & \text{if } Cat\,(I2) = CAT\,(I1) = C \end{cases} \quad (4)$$

Degree of Similarity Between Examples and Context-window

Definition: Assuming Root(w) is the corresponding word of a node whose serial number is w, and Cat_Root(w) is the part of speech of Root(w). Given a context-window W=$\{w1,w2,...,wm$, If the example EG takes the form of ((eg1,eg2,...,egk)(egk+1,...,egn)), the first element of the list is the front set of the example FRONT–EG(EG)={eg1,eg2,...,egk}, the second the back set BACK–EG(EG) ={egk+1,...,egn}; When EG={eg1, eg2,...,egn}, then the FRONT–EG(EG) is null and the BACK–EG(EG) =EG. The degree of similarity of the elements of EG under W is:

Sem-Eg-items(EG,W) = $\underset{\substack{e\in \text{BACK–EG(EG)} w\in W,\\ w \text{ has not the label SUBJ}}}{\cup}$ MAX SIM-WORD(e, Root(w),Cat(w))

$\qquad\qquad\qquad$ $\underset{\substack{e\in \text{FRONT–EG(EG)} w\in W,\\ w \text{ has the label SUBJ}}}{\cup}$ MAX SIM-WORD(e, Root(w),Cat(w))   (5)

It is desirable that the element with the highest score in the set Sem-Eg-items (EG,W) be able to have a greater influence on the evaluation of the examples. For this purpose, we use weights to get the degree of similarity between EG and the context-window:

Sem-Eg-Win(EG,W) = ($\underset{1<k<10}{\Sigma}$ (Weight(i)x $\underset{\substack{x\in \text{SEM-Eg-items(EG,W)},\\ (k-1)x10<x<kx10}}{\Sigma x}$))/ | EG | (6)

The weight function Weight(k)(1<k<10) in the formula should be a monotonously-increasing function and its value should increase sharply in the section [0,1] to amplify the effects of higher scores.

Evaluation of the Determining Word

Given the context-window of a word, W, and the example set of *ith* entry of the word, EGs(i), we can calculate the score of the word entry:

Value – Entry(I, W) = $\underset{EG \in Egs(I)}{\Sigma}$ Sim – Eg – Win(EG, W)   (7)

Assuming that the serial number of the context-window of the node N is not empty, which means the corresponding word P of the node N can act as the determining word of some other word, and that the entry set of N is Entries-Node(N), then the final score of the node, Value-Node (N) is:

Value – Node(N) = $\underset{I \in \text{Entries – Node(N)}}{\text{MAX}}$ Value – Engry(I, WIN(N))   (8)

Evaluation of Non-determining Words

Assuming that R is the notional word set of a parsing tree, *r* R, and Win(*r*) is empty, If the set Master(r)={$x|x$_R, x determinates r} is empty, that means that r cannot act as the determining or determined words of any other notional words in R. Then r has no any relationship with other words, and has no effect on the structure disambiguation. In the algorithms we do not process this kind of notional words. Otherwise, if Master(r) is not empty, we can get the score of r according to the matching relation between r and the element of Master(r). Here will be at least one of the notional word M in Master(r), the example set of whose determined entry has been compared with r. We can take the highest degree of similarity as the score of r under M, and the word entry that gets this highest score as the selected entry. The final score of r is the maximum of the score of r under Master(r). If r gets the highest score under M'_Master(r), then the final selected entry is the entry of r under M'.

Evaluation of the Parsing Tree and the Determination of the Parsing Result

Given a dependency tree with m node T={N1,N2,…,Nm}, where Ni is one of the nodes in the tree, the score of T is calculated as follows:

$$\text{Value} - \text{Tree}(T) = \sum_{n \in T} \text{Value} - \text{Node}(n) \qquad (9)$$

When a input sentence analysed has k different dependency tree, $t1$, $t2$, …, tk, and $k>1$, The tree with the highest score will be taken as the parse result:

Value-Result = Max{Value-Tree(t1),Value-Tree(t2),…,Value-Tree (tk)} (10)

Example of the Disambiguation Algorithm

The following is a synopsis of the disambiguation procedure:

Input phrase: 維修圖書館的空調
Syntax parse result:
   R1: ((維修/v 圖書館/n) 的 空調/n)/np，
   R2: (維修/v (圖書館/n 的 空調/n))/vp

The dependency tree of R1 and R2 is illustrated in Figure 1.

The determining words and their context-windows are listed in Table 1. The word entries and their example information come from *Hownet*.

Word	DEF of the entry	Example of the entry
維修	repair | 修理	工人~電器，爸爸在~電視，~電冰箱
圖書館	InstitutePlace | 場所, @read | 讀, @borrow | 借入, #readings | 讀物	
空調	tool | 用具, * adjust | 調整, # temperature | 溫度	
電器	tool | 用具, generic | 統稱, # electricity | 電	
電視	image | 圖像, shows | 表演物	
	tool | 用具, * look | 看, # image | 圖像, # shows | 表演物	
工人	human | 人, # occupation | 職位, industrial | 工	

According to the algorithm mentioned above, the degree of similarity between the example set of the entry of every notional word and its context-window is calculated. The scores of every notional word and the whole dependency tree are obtained from the calculation:

	維修	圖書館	空調	TOTAL
R1	0.45	39.88	23.41	63.74
R2	88.46	0	14.39	102.85

Obviously, the final parse result is R2.

The Results of the Experiment

We chose 800 sentences from volumes 1–4 of the Chinese elementary school textbooks as the experimental corpus. First, the sentences are parsed using the RCAS MT system developed by our project team. Then the parse result is taken as the input of our disambiguation system. The parse result is evaluated manually using the correctness rate of the disambiguation CR, which is defined as:

CR = the number of sentences that can get the correct parsing result after disambiguating/the total number of sentences parsed (10)

At present the CR is 91.4%. This result shows that the knowledge-based method is very effective as regards to disambiguation.

In the experiment, we also examined the relationship between the time spent and the length of the sentence, as well as the average number of a word entry (see Table 2).

Table 2. The Average Evaluating Time under Different Length and Entry Number (Unit Time)

Length \ Entry number	1–5	6–9	10–13	14–17	18–21
1–3	17	32	44	56	84
4–6	70	101	163	217	283
7–9	200	362	562	767	1001

From the above table we find:

(1) The number of entries of the words in the sentence has greater effect on the evaluating time. This is because when evaluating a determining word we should consider all the matching relations between the word entry and of those in the context-window. So if the length of sentence is fixed, the degree of complexity has direct relationship with the square of the average number of entry of words in the sentence.

(2) When the entry number is fixed, the time has a direct relationship with the length of the sentence. The main reason is that the length of the context-window of every determining node in the dependency tree does not exceed the total number of its parents and children.

At present *Hownet* has not yet provided all the examples for every word entry. In the experiment we specified examples for some words manually, which can lead to some skewing. In addition, the number of sentences used in the experiment is also limited. We hope that in the future *Hownet* can provide all the examples so we can carry out an open experiment on a larger corpus and obtain more reliable experimental data.

Conclusion

In this paper we present a method of Chinese syntactic structure and word sense disambiguation. The method selects the best candidate from the multi-parsing trees by comparing their degree of syntactic and semantic collocation between words in the structure analysed. The knowledge comes from the examples in each word entry in *Hownet*, a knowledge base especially designed for NLP. Our experiment shows that the method is very effective for the purpose. And it is obviously more tolerant and much better than the traditional clear-cut method of YES or NO.

References

Dong, Zhendong 董振東 and Dong Qiang 董強 (n.d). *Hownet* 〈知網〉. http://www. how-net.com

Feng, Zhiwei 馮志偉 (1995). 〈歧義消解策略初探〉 (Some Initial Methods of Disambiguation).《計算語言學進展與應用》(*Developments and Applications in Computational Linguistics*). Beijing: Tsinghua University Press.

Lu, Shuxiang 呂叔湘 (1990). 〈中國文法要略〉 (Outline of Chinese Grammar), in 《呂叔湘文集》(*Collected Works of Lu Shuxiang*), Vol. 1. Beijing: The Commercial Press.

Shao, Jingmin 邵敬敏 (1986). 〈關於歧義結構的研討〉 (Research on Disambiguation).《語文導報》(*Language Bulletin*).

Yuan, Chunfa 苑春法 and Wong Kam Fai 黃錦輝, *et al.* (1999). 〈基於語義知識的漢語句法結構排歧〉 (On Semantic Knowledge in Chinese Syntax Disambiguation).《中文信息學報》(*Journal of Chinese Information Processing*), Vol. 13, No. 1.

Zhang, Jie 張傑, Zhang Yue 張躍 and Yao Tianshun 姚天順 (1999). 〈面向資料的英漢機譯系統中關於組合基於實例的技術〉 (The Example-based Technique in an English-Chinese Machine Translation System).《東北大學學報》(*Northeastern University Bulletin*), Vol. 20, No. 3.

Zhang, Min 張民 and Li Shengdeng 李生等 (1995). 〈一種漢語句子間相似度的度量演算法及其實現〉 (Calculation of Measuring the Syntactical Similarity in Chinese Sentences).《計算語言學進展與應用》(*Developments and Applications in Computational Linguistics*). Beijing: Tsinghua University Press.

Zhang, Yumin 張玉敏 (2000).〈機器翻譯中句法語義相結合的漢語分析研究〉 (Analytic Study of Chinese Syntax and Semantic Meaning in Machine Translation). Master's Dissertation, Xiamen University.

Zhou, Huiping 周會平 (1999). 〈基於中間語言漢英的翻譯系統ICENT的研究與實現〉 (On the Research and Application of Intralingual Language in the ICENT Chinese-English Translation System). Doctorial Dissertation, Postgraduate School, National Defence Science and Technology University.

Example-based Machine Translation: A New Paradigm

Kit Chunyu, Pan Haihua and Jonathan J. Webster
Department of Chinese, Translation and Linguistics
City University of Hong Kong

Introduction — Why EBMT?

Machine Translation (MT) is aimed to enable a computer to transfer natural language utterances in either text or speech from one language into another while preserving the meaning and interpretation. MT technology has gone through several paradigms from its very beginning in the past half century, including word-to-word *direct* translation, rule-based *transfer* approach, *interlingua* approach and *Knowledge-based* Machine Translation (KBMT). The direct translation approach relies too much on dictionary look-up. The transfer approach incorporates language analysis and representation at various linguistic levels, but cannot find adequate knowledge to resolve ambiguities involved in the language analysis, transfer and generation. The inter lingua approach rests on the assumption that all languages share a common underlying representation, but such a goal appears unreachable. The *knowledge-based* approach attempts, mostly in the fashion of knowledge engineering (KE) in traditional symbolic artificial intelligence (AI), to acquire and encode various kinds of knowledge (e.g., encyclopaedic knowledge) for the purpose of disambiguation, but the source of knowledge remains a serious problem. Most recent MT research falls into statistical MT and example-based MT. It is argued in Somers (1998; 2000a) that statistical MT is also an example-based approach.

In general, the translation process in traditional MT involves three subtasks, namely, analysis, transfer and generation, as depicted in the MT pyramid. Analysis deals with the transformation of the source utterances into a predefined format of internal representation, through morphological processing, part-of-speech (POS) tagging, syntactic parsing, semantic analysis, etc. Transfer works to convert the representation of the source language into that of the target language. Except for the inter lingua approach, there used to be different representations for different languages. Generation, or synthesis, is concerned with the derivation of target utterances from the representation, observing necessary syntactic, semantic and pragmatic constraints. The generation process could be thought of as the reverse process of the analysis, but actually is quite different in nature. Each step in these processes involves, inevitably, many ambiguities, and each kind of ambiguity needs a large volume of knowledge for comprehensive disambiguation. For example, a bilingual dictionary is far less enough for word sense disambiguation (WSD) to determine what a word — which usually has many senses — exactly means within a particular context.

As in other language engineering tasks, MT needs a vast amount of knowledge for disambiguation, which is a kind of decision-making during the above translation processes. How to acquire adequate knowledge for reliable disambiguation is one of the most critical issues in current MT technology. The symbolic KE approach may have been a possible solution, if it had matured enough to successfully handle several significant problems in large-scale practical KE. For a practical language engineering task such as MT, the problems include how to acquire enough knowledge, how to represent and encode different kinds of knowledge, how to maintain the knowledge base and resolve the scale-up problem, etc. In the past decades it was the practice to have experts (e.g., syntacticians) write down expert knowledge (e.g., grammar rules) in some rule-based format. This achieved significant but limited success. The inadequacy of manually encoded knowledge remains a problem — there are always so many practical ambiguities that experts cannot foresee during the construction of a knowledge base. The maintenance and scale-up problems also emerge when the knowledge base becomes larger and larger. For example, changing a single rule might cause unpredictable conflicts with other rules and, consequently, lead to a crash of the entire rule system.

Thus, there is a necessity to look for an alternative approach to knowledge engineering for MT that can automatically acquire practical knowledge from, and also adapt itself towards, real language data. EBMT is considered one of the current attempts towards this goal. The basis for EBMT is the existence of a large volume of translated texts (i.e., parallel bilingual texts), which have been translated by professionals with not only language proficiency but also specialist expertise. In this sense, bilingual texts encode knowledge that can be extracted to facilitate the automatic translation. Technically speaking, EBMT is about how to "decode" knowledge from bilingual texts, where the knowledge seems to have no overt formal representation or any encoding scheme. Instead, such knowledge is encoded in a way as straightforwardly as text coupling: a piece of text in one language matches a piece of text in another language.

In this article, we will give an overview of the EBMT technology. In the next section we will review the history of EBMT, with a focus on the main ideas. Since a comprehensive review of EBMT can be found in Somers (2000a), we will focus on the discussion of our viewpoints of the EBMT framework. Then we will define the notion of example and examine the major issues involved in EBMT, covering mainly the four major stages of EBMT, namely, example acquisition, example base management, example application and target sentence generation. Some of our current work in lexical-based text alignment for example acquisition is also discussed, highlighting the formulation of a similarity measure and alignment algorithm, before concluding our discussion in the last section.

History

EBMT was most notably attributed to Nagao and his famous "translation by analogy" paper in 1984. Relevant ideas, however, can be dated back to 1970's, if not earlier. In view of this, it appears more appropriate to think of EBMT as having multiple origins, among which Kay's (1997) Translation Memory (TM) and Nagao's (1984) translation by analogy, in our viewpoint, are the most influential ones.

According to Melby (1995), the Brigham Young University MT Group incorporated a similar idea into the ALPS system, one of the earliest commercial MT systems, as a "Repetition Processing" tool. Arthen (1978) proposed "a programme which would enable the word processor to 'remember' whether any part of a new text typed into it had already been translated, and to fetch this part together with the translation." He foresaw that a translator could benefit greatly from online access to similar translated texts while translating a new text, if "the system would check this text against the earlier texts stored together with other ... languages" and provide a simple operation as "cut and stick." Kay (1980) proposed the translation amanuensis, i.e., a bilingual text editor as a workbench for translation, as a less ambitious but more realistic starting point for the MT technology, and put forward the idea of having the machine memorize existing translations to facilitate ongoing translation in his famous "proper place" paper on the relation of men and machines in language translation: "... the translator ... issuing a command causing the system to display anything in the store that might be relevant ... he can examine past and future fragments of text that contain similar material." This idea, having been suggested by others since the early 1970s (Hutchins, 1998), was later called TM. In general, TM can be understood as a restricted form of EBMT.

The essence of Nagao's EBMT, or "machine translation by the analogy principle," is more comprehensive, however, than TM. It attempts to mimic the cognitive processes of human translators for the purpose of automating the translation process. There are three main tasks in EBMT, as identified by Nagao, including

- Matching fragments against existing examples,
- Identifying the corresponding translation fragments, and then
- Recombining them to give the target text.

Among these three, TM shares the first. In Nagao's own words: "Man does not translate ... by doing deep linguistic analysis. Rather, man does translation, first by properly decomposing an input sentence into certain fragmental phrases ..., then by translating these phrases into other language phrases, and finally by properly composing these fragmental translations into one long sentence." Nagao (1984) also proposed a matching technique based on measuring the semantic proximity of words. A similar idea was proposed independently by the DLT group in Utrecht at about the same time, according to Pappegaaij *et al.* (1986) and Schubert (1986).

In the early 1990s, many MT researchers were attracted to two new paradigms of MT, one being statistical-based machine translation (Brown *et al.* 1990, Brown *et al.* 1993) and the other being Nagao's EBMT. In a review paper on EBMT, Somers (1998) notes that "The statistical approach is clearly example-based in that it depends on a bilingual corpus, but the matching and recombination stages that characterise EBMT are implemented in quite a different way in these approaches."

The EBMT approach became popular soon after some positive results were published in a number of papers demonstrating its plausibility. Sato and Nagao (1990) investigated the problem of example selection by approximate (or inexact) matching of input sentences and example sentences, using a similarity measure based on the syntactic similarity of dependency tree structures of a sentence pair in question and on the word distance (i.e., a similarity value) of corresponding words, which were pre-defined in a thesaurus. Sumita *et al.* (1990) looked into example-based translation of Japanese noun phrases of the pattern [N1 *no* N2] into English as [N2 *prep* N1] or [N1 N2], based on a distance measure for the input phrase and example phrase, calculated as a linear weighted sum of the distances of the three subparts, each of which is predefined in a thesaurus. However, the serious difficulty in constructing a large thesaurus with reliable similarity values between so many word pairs would prevent these two theoretically interesting approaches from having any practical application except for very limited domains. The problem of sparse data also seems to prevent these two approaches, one limited to using sentence level examples and the other to phrase examples, from further success in practice.

By around 1993, EBMT had become an established research field of MT and many example-based techniques were applied to various MT tasks. Sato (1993) attempted the example-based translation of computer technical terms with respect to the focus term and its surrounding contexts and reported an overall accuracy of 96%, with an accuracy of 92% for unknown terms. Sumita *et al.* (1993) tackled the prepositional phrase attachment problem in translation with example-based techniques, using the same similarity measure as in Sumita *et al.* (1990) The research focused on nine most frequently used English prepositions and the positive results suggest that this approach can be generalised to many other prepositions and other types of structural ambiguity. Nirenburg *et al.* (1993) takes a closer look into the problem of matching an input sentence against possible examples, based on a distance measure defined in terms of necessary keystrokes in editing operations (e.g., deletion = 3 strokes, substitution = 3 strokes) required to convert an example into the input sentence. The experiments using *Wall Street Journal* data suggest that example matching at the sentence level is undesirable and partitioning a sentence into sub-sentential strings is necessary, in order to achieve an acceptable example matching result. Nirenburg *et al.* (1994) and Brown (1996) reported their progress in integrating EBMT with KBMT.

Unfortunately, however, there has been little research on EBMT related to Chinese ever reported, although a number of researchers have explored the lexical acquisition at word and phrase levels by text alignment techniques. For example, Wu and Xia (1995) report successful results from English-Chinese text alignment at various levels using statistical-based methods; Fung and McKeown (1997) carried out English-Chinese terminology translation using a noisy parallel corpus.

What Is an Example?

In the early research of EBMT, e.g., in the early 1990s, many researchers tended to focus on examples at the sentence level. Since sentential examples that can exactly match input sentences are rare, researchers consider the possibilities of looking for similar examples or translation templates for translating new sentences. But the effectiveness seems problematic.

In general, an example is a pair (or couple) of texts in two languages that are a translation of each other. The texts can be of any size at any linguistic level: word, phrase, clause, sentence, and event paragraph. More flexibly, an example need not match a linguistically meaningful structure or constituent, that is, an example can simply be a pair of text chunks of an arbitrary length. However, we know that longer examples have a smaller chance of showing up in incoming texts. It is not strange that many sentential examples have no chance to be hit again during the phase of example application in EBMT. Thus, we need to pay attention to the usefulness of an example when considering what examples need to be acquired and put in the Example Base (EB) — the collection of examples.

A critical issue that needs to be examined closely in this context is the number of examples over a large-scale bilingual corpus, which can be unlimited in practice. Notice that an example can be further decomposed, in more than one possible way, into sub-structures or shorter examples, and that examples can overlap with each other. Therefore, the example number can be exponentially large in respect to the corpus size, if we collect all possible examples from a bilingual corpus. Consequently, the impracticality and implausibility of EBMT might arise, because any fragment of a sentence can be an example. We know that a language is well-known for utilising limited resources (e.g., lexicon and grammar) to produce an unlimited number of utterances. Thus, it is an interesting issue to examine the practicality of EBMT in terms of the correlation of example number and corpus size. In practice, how to control an EB to a reasonable size becomes vitally critical. For this purpose, we need to determine what examples should be filtered out and which ones should be maintained in the EB, not only for the matter of efficiency but, more importantly, for practicality. Although we have not come up with a clear strategy for example control, it is nevertheless an important issue in example base management.

The relation of bilingual dictionary and EB should also be carefully examined. Conventionally, a bilingual dictionary is a collection of lexical entries in one language and gives many possible translations in another language for each word. We can think of a bilingual dictionary as a restricted EB, containing a collection of examples restricted at the word level. In return, an EB can be regarded as an extended bilingual dictionary. One might point out the fact that translating a word into another language following a bilingual dictionary is so uncertain, but translating a multiple word fragment of utterance in terms of an example from the EB is, in contrast, more sure. But this does not necessarily mean the dictionary and the EB do not share an intrinsic property for translation, namely, they provide choice of translation for a fragment, either single or multiple word, in an utterance. We can conceive — actually, we have observed — that just like lexical entries in a dictionary, examples in the EB are also not limited to one-to-one mapping, because many utterance fragments can have more than one translation. All choices need to be collected in the EB, highly similar to that in a dictionary, although the choices are significantly fewer.

Hence, an empirical MT approach like EBMT can be understood as to tackle the following problem: given some observed translation as a set of utterance fragments (either word, phrases, sentences or even some text chunks) with their possible choices of translation in another language, find a reasonably good, if not the best, translation for the next input utterance.

The Four Stages of EBMT

In general, there are four stages of work in EBMT, namely, example acquisition, example base management, example application and target sentence synthesis. Example acquisition is about how to acquire examples from parallel bilingual corpus (i.e., existing translation), and example base management is about how examples are stored and maintained. The example application concerns itself with how examples are used to facilitate translation, which involves the decomposition of an input sentence into examples and the conversion of source texts into target texts in terms of existing translation. The sentence synthesis is to compose a target sentence by putting the converted examples into a smoothly readable order, aiming at enhancing the readability of the target sentence after conversion.

Example Acquisition

There are various resources from which we can collect examples. For example, from bilingual dictionaries we can collect examples at the word level. It is in this sense that an example base can be thought of as an extended bilingual dictionary that covers examples beyond the word level. These multiple-word examples have to be acquired from bilingual corpora, most of which are usually aligned at the article or even the paragraph level.

Text alignment seems to be a necessary step towards example acquisition at various levels. Manual alignment by experts can, of course, produce quite reliable examples, but the price for precision is a problem, and the speed is far less enough to handle a corpus of millions or tens of millions of words for practical applications. Thus, we have to resort to automatic text alignment technology.

The approaches to text alignment can be categorized into two types, namely, resource-poor and resource-rich approaches. The resource-poor approach mostly focuses on sentence alignment and relies mainly on sentence length statistics, co-occurrence statistics and some limited lexical information, as illustrated in Kay and Roscheisen (1988, 1993), Gale and Church (1991), Brown *et al.* (1991), Chen (1993), Church (1993), among many others. In addition to a collection of examples, a bilingual lexicon is also expected to be inferred from a parallel corpus via alignment, known as word alignment. In contrast, the resource-rich approaches make use of whatever available and useful, in particular, bilingual lexicon and glossary, to facilitate the alignment. We will have some more discussion on these approaches later.

Examples to be learned are not limited to the sentence level. Rather, we are more interested in learning examples at sub-sentential levels, including words, idioms and collocations, multi-word terminology, and phrases. The alignment at the levels of phrase, collocation and word is more critical, because they are the examples whose target language parts are so productive in the composition of the translation output. Some critical techniques have been developed in previous research, e.g., word alignment in Dagan *et al.* (1993), partial parsing for EBMT in Furuse and Iida (1994), and bilingual parsing with the stochastic inversion transduction grammars in Wu (1995a, 1995b, 1997).

A comprehensive review on text alignment technology can be found in Wu (2000). The discussion below will focus on the work of text alignment with lexical resources that we are undertaking for the purpose of example acquisition.

Similarity Measure

A similarity measure is required in text alignment to give an indication to which pair of texts in a bilingual corpus is more likely to be a match. For example, given a bilingual corpus that has an average sentence length ratio r, it is reasonable to believe that a pair of sentences whose length ratio is close to this r are more likely to match each other than two sentences whose length ratio is far different. The observation underlying this belief is that a short sentence in one language is usually translated into a short sentence in another language, and so are the long sentences. Also, sentence position is also useful information to exploit. Sentence pairs whose two sentences are in positions far away from each other in a parallel bilingual corpus have no doubt a lower chance to match each other than the ones that are closer to each other.

In addition to the factors of sentence length and position, a resource-rich approach may also take into account the matched pairs of dictionary items and glossary (or terminological) items. Accordingly, a formula for scoring the similarity of a sentence (or clause) pair $<s_p, s_q>$ in a given bilingual corpus can be empirically formulated as below:

$$sim(s_p, s_q) = \frac{\sum_{i=1}^{m} f(d_i) + w \cdot \sum_{j=1}^{n} f(g_i)}{\mid r \cdot l(s_p) - l(s_q) \mid \cdot \mid i - j \mid} \tag{1}$$

where p and q are sentence (or clause) positions, d_i and g_i are matched dictionary and glossary items, respectively, $f(\cdot)$ is an evaluation function for the significance of a matched item in the similarity measure, is a weight indicating how many times a matched glossary item is as important as a matched dictionary item in the similarity measure, and $l(\cdot)$ is the length of a given text. The simplest evaluation function is $f(\cdot) = l(\cdot)$, which means that we take the length, say, in characters, of a matched item as the measure for its significance in the similarity measure. Since a pair of matched items involves two parts in two different languages, the way to combine the length of the two into one looks less straightforward than simply summing up their lengths. An equivalence ratio of string length in characters in different languages needs to be taken into account. However, to simplify the issue, it would not be a poor choice to consider only the string length in one language.

The coefficient is to be determined for individual corpus through experiments. According to our experimental results on clause alignment with this similarity measure on BLIS, the English-Chinese bilingual corpus of Hong Kong legal texts of about 20 million words, the optimal value for w is 8, which indicates that a matched glossary item is about eight times as important as a matched dictionary item in general.

Also, since matched glossary and dictionary items both give an indication of a good alignment overwhelmingly stronger than the clause length and position information, we have experimental results to show that omitting these two factors in the above similarity measure leads to no significant difference in our resource-rich approach to clause alignment for the parallel corpus of Hong Kong legal texts.

Alignment Algorithm

When a similarity measure is available, we can calculate the similarity for all sentence pairs in a given bilingual corpus. This calculation produces a similarity matrix. For example, below is an illustration with a few sentence pairs, where $sim(c_2, e_1) = 0.6$. For simplicity, we may denote a score $sim(c_i, e_j)$ as a_{ij}; accordingly, $a_{11} = 1.2$ and $a_{12} = 0.6$.

	e_1	e_2	e_3	$\cdots$
c_1	1.2	2.3	0.4	$\cdots$
c_2	0.6	0.9	2.5	$\cdots$
c_3	0.7	0.8	7.5	$\cdots$
$\cdots$	$\cdots$	$\cdots$	$\cdots$	$\cdots$

Once a similarity matrix is available, the alignment algorithm to pick up a set of scores in the matrix that covers all sentences in the corpus can be rather straightforward. Remember that the rule of thumb for the alignment algorithm is that every sentence in each language tends to match a sentence in another language with the highest similarity score. Following this, we can have an alignment algorithm as follows with regard to a given similarity matrix:

(1) Pick the greatest score in each row;
(2) Pick the greatest score in each column;
(3) Derive the union of the two sets obtained from (1) and (2).

With the above matrix as an example, the algorithm picks $\{a_{12} = 2.3, a_{23} = 2.5, a_{33}\}$, in step (1) and $\{a_{11} = 1.2, a_{12} = 2.3, a_{33} = 7.5\}$ in step (2), and the union of the two sets is $\{a_{11} = 1.2, a_{12} = 2.3, a_{33} = 2.5, a_{33} = 7.5\}$, as marked in bold face below. Consequently, the alignment gives the result $\{<[c_1] : [e_1, e_2]>, <[c_2, c_3] : [e_3]>\}$, with the former maps one sentence (or clause) to two and the latter two to one.

	e_1	e_2	e_3	$\cdots$
c_1	1.2	2.3	0.4	$\cdots$
c_2	0.6	0.9	2.5	$\cdots$
c_3	0.7	0.8	7.5	$\cdots$
$\cdots$	$\cdots$	$\cdots$	$\cdots$	$\cdots$

We can see that this simple algorithm derives not only one-to-one but also one-to-many alignments, and that many-to-many alignments are also conceivable. In practice, however, the interpretation of the set of scores chosen from a similarity matrix is not as straightforward as the above example, because not a few many-to-many alignments may be involved, although an overwhelming number of alignments are one-to-one. According to our observation, the many-to-many alignments are not rare in practice. An inadequate capacity of handling such alignments would no doubt have a significant negative effect on the alignment performance over a large-scale corpus.

Evaluation of Alignment Results

Conventionally, precision and recall in terms of the proportion of correctly aligned sentences (or clauses) are used to measure the alignment performance. The precision is the proportion of correctly aligned pairs in all aligned pairs, and the recall is the proportion of correctly aligned pairs in all correct pairs. For example, given a bilingual corpus of M pairs and an alignment output of N pairs among which A pairs are correct, the precision $P = \frac{A}{N}$ and the recall $R = \frac{A}{M}$. A is, in a sense, the overlap of M and N.

However, since sentences and clauses are of various lengths, such percentage measures give only a rough measure about the performance. We cannot say that the correct alignment of a sentence pair of twenty words is of the same significance as that of a pair of ten words. Thus, there seems to be a necessity to develop a more comprehensive measurement for the performance of text alignment with some finer-grained measures. For example, the proportion of words and characters in correctly aligned sentences may also give an indication of the performance in addition to the precision and recall. There is no doubt that the combination of these measures gives a more comprehensive and reliable evaluation of the performance.

In the evaluation, it is also meaningful to examine the performance against the sentence lengths and resources used. Some approaches may have a better performance on long sentences and some on short sentences. Some resources play a more important role than others.

Straightforwardly, a more comprehensive evaluation is needed for an alignment approach that carries out alignment at various linguistic levels simultaneously. For example, if an algorithm is to derive text pairs at the levels of clause, phrase and word, we have to look into the precision and recall at all these levels for the purpose of evaluation.

Example Base

Once the text alignment phase is carried out from top to bottom at various structure levels, we actually have had an entire collection of examples in the aligned bilingual corpus. Conceptually speaking, all examples can be extracted from the aligned corpus and technically we need an example base (EB) for convenient storage and retrieval of examples. Whether it is necessary to extract all examples from the corpus is also a non-trivial issue, because an aligned bilingual corpus is itself storage for the examples. More importantly, the EB is not merely a place to pile up the examples. Rather, it needs to play the role of a language model, with the examples as structural parameters each associated with a probability estimation, for the purpose of facilitating later stage processes of EBMT, e.g., the decomposition of a source sentence into existing examples and the composition of a target sentence from a sequence of its component examples.

Example Extraction

Extracting examples from a well-aligned bilingual text is not a trivial task. To be given a set of scores chosen from a similarity matrix to determine the alignment is also far from trivial. Let us suppose that a sentence pair is word-aligned, then what pairs of word sequence should be extracted as examples? For example, given the following sequences of aligned words, what examples are going to be extracted?

$$
\begin{array}{ccccc}
\ldots & a & b & c & d & \cdots \\
& | & | & | & | & \\
\ldots & A & B & C & D & \cdots
\end{array}
$$

Word pairs, of course. And then, two-word pairs, like <ab : *AB*>; three-word pairs <abc : *ABC*>; and so on?

A serious problem with this approach is the control of example number. The number of such examples, among which many are useless, would be too large on a corpus of millions of utterances. Given n word pairs, the total number of examples as the above is in the order of $O(n^2)$. For a corpus of m utterances with the average length n, the number of examples is in the order $O(mn^2)$. Our computers seem to have adequate memory space to handle this complexity before m goes too large. However, a huge m is commonly encountered in current empirical approaches to human language technology, e.g., $m = 1,000,000$. Assume the average utterance length $n = 10{\sim}15$ and each example needs only 10 bytes, the space needed for the examples would be $10mn^2 = 1.0{\sim}2.25G$ bytes. What if we have a corpus ten times larger? Notice that a larger corpus is always welcome, because it contains more knowledge that can facilitate MT. Thus, instead of relying on the computer's memory capacity, we pay more attention to what examples — if extracted — should be filtered out, in order to enhance the efficiency. Structural analysis may help us to extract only sentential constituents, to prevent too many meaningless word sequences from entering the EB. But obviously, a rigorous measure on the usefulness of an example seems necessary.

In addition to examples with continuous word sequences as the above, are there any discontinuous sequences that should also be extracted? For example, "take [something] into account" may map to something like "consider [something]" in another language. We can see that many template-like examples are quite common and useful. This appears to be another direction of research concerning example extraction: acquisition of template or patterns from examples, where some machine learning techniques may be involved.

Example Base Management

The example base is a crucial component in an EBMT system. It handles the storage, edition (including addition, deletion and modification) and retrieval of examples, to support the translation process, be it fully automatic or human-aided. Thus, an efficient EB must be capable of handling a massive volume of examples at an adequately high speed.

The format for literal example in the EB can be simple: a sequence of words in both the source and the target language is appropriate, although some more sophisticated language mark-up technology such as Extensible Markup Language (XML) or Resource Description Framework (RDF) can undoubtedly be helpful. Also, an efficient strategy for searching through an EB 139 Translation and Information Technology of tens or even thousands of millions of example entries is necessary. In this direction, it is certainly beneficial to incorporate the machine-readable dictionary (MRD) technology (Evans and Kilgarriff, 1995) into the EB management.

If example patterns (or templates) and some rule-based examples bearing context-sensitive information about the preferable translation under a particular context are also considered, heterogeneous formats instead of a uniform format need to be employed in the EB for example representation and encoding. Accordingly, it is necessary to develop a more comprehensive and sophisticated EB management technology.

Example Base as a Language Model

The EB may also play the role of a language model that needs to be utilised in later phases of EBMT to choose a suitable sequence of available examples for the translation of an input sentence, and determine the word and/or chunk order in the target sentence for better readability. The former step is known as source sentence decomposition and the latter target sentence synthesis.

In a language model, structural items (e.g., the examples) are usually associated with some statistical parameters, e.g., frequencies or probabilities. In the case of n-gram model, the parameters are attached to item sequences. Frequency information is expected to incorporate not only the counts in a given bilingual corpus but also usage frequency of the examples by users. For this purpose, the frequency information needs to be updated dynamically during translation practice.

Example Application

The phase of example application is about how to make use of existing examples to do translation. The essence of EBMT is this: whenever you see a piece of input text that has been translated before, simply use the existing translation. EBMT is about how to make use of existing translation to translate new texts following this principle. It is rather simple to translate an input sentence that has already been translated before: copy its translation to the output.

However, to retranslate a seen sentence is rare in practice. And we also observe that translators spend most of their time dealing with new sentences with translated fragments. Hence, a more critical and interesting issue in EBMT is how to translate a new sentence with fragments that are already translated.

Thus, the first task in the stage of example application is to decompose, or segment, an input sentence into a sequence of seen fragments, namely, examples. Usually, there can be more than one way to decompose a sentence. Among these possibilities it is reasonable to choose the best for the next phase of translation. However, by what criterion can we say one choice is better than the other?

There can be different criteria about the best ways to decompose a sentence. One of the choices is probability: we segment an input sentence s into a sequence $d(s)$ of examples $e_1 e_2 ... e_n$ that is most probable in terms of a given language model. Formally put, this idea can be formulated as

$$d(s) = \arg\max_{e_1 e_2 ... e_n = s} p(e_1 e_2 ... e_n) \qquad (2)$$

where the probability $p(\cdot)$ can be computed in terms of some language model. For example, if we use a multi-gram model (of words), we have

$$p(e_1 e_2 ... e_n) = \prod_n p(e_i) \qquad (3)$$

The probability $p(e_i)$ can be approximated, in the simplest way, by its relative frequency in the example base, namely,

$$p(e_i) = \frac{f(e_i)}{N} \qquad (4)$$

where N is the frequency sum of all examples in the given corpus. This is the simplest approximation. We can resort to language modelling technology for a more accurate estimation for this probability.

When the sentence decomposition is done, the next task in example application is to convert the resulting fragments from the source language into the target language. It looks like a trivial task, if there is no more than one choice of translation for a fragment in question. However, in the case of multiple translations available for a fragment, the problem appears highly similar to the one known as *word sense disambiguation* (WSD) (Yarowsky, 2000), in which the now fully-fledged pos-tagging technology may find a critical role to play. (Wilks and Stenvenson, 1998) We may hope that this technology can alleviate the problem to a great extent.

Another appropriate solution would be that we postpone the problem to the next phase, namely, target sentence synthesis, where we can pick a sequence of target fragments that has the highest *readability*, or *smoothness* of reading, among all possible sequences. This smoothness is, as we would suggest, to be measured by some probability over the sentence in question with regard to a language model, say, a multi-gram model, for the target language.

Sentence Synthesis and Smoothing

After sentence decomposition and example transfer, we have a sequence of translated fragments. The next task is to combine these translated chunks into a well-formed highly readable sentence in the target language. This is recognized as the most difficult step in EBMT but "has received considerably less attention." (Somers, 1998)

Since different languages have different syntax to govern the sentential structures and word order, it won't work in most cases if we simply chain up the translated fragments in the same order as in the source language. The most critical point in this stage is to adjust the fragment order to form a smoothly readable sentence in the target language. In this sense, the sentence synthesis for EBMT is a job of smoothing for the enhancement of readability.

In general, language generation needs practical strategies for the sentence composition that consider not only the internal structure and external context of the input sentence (Collins and Cunningham, 1997), but also stylistic and discourse issues with respect to culture fidelity and available text generation techniques for MT. A set of grammar rules is surely not adequate, no matter how comprehensive the grammar would be. Remember that EBMT takes an empirical case-based knowledge engineering approach to MT. Thus, it is conceivable that a more preferable strategy for sentence synthesis is a probabilistic approach with a language model for the target language, which computes the probability for any ordering of a given set of chunks of words. Among all possible orderings, we choose the most probable one.

Given a set of translated chunks $\{c_1, c_2, ..., c_n\}$, we look for the best ordering of them, as formulated below:

$$s(c_1, c_2, ..., c_n) = \underset{c_1' c_2' ... c_n' \in O(c_1, c_2, ..., c_n)}{\arg\max} p(c_1' c_2' ... c_n') \qquad (5)$$

where $O(\cdot)$ denotes the set of all possible orderings of a given set of chunks. Notice that this time a linear multi-gram model does NOT work, because the chunk order has no effect on the probability of the chunk sequence, as shown in (3).

What we actually need here is a language model that is sensitive to word order in an ordering of the given chunks. The simplest choice is a fixed-order n-gram model, e.g., a bi-gram or tri-gram model. Some more sophisticated models are of course applicable, e.g., a probabilistic context-free grammar, but some more complicated NLP processing tasks such as parsing may be involved.

For example, if a tri-gram model is chosen, the probability $p(\cdot)$ in (5) can be formulated as below accordingly:

$$p(c_1' c_2' ... c_n') = \prod_{w_{i-2} w_{i-1} w_i \in c_1' c_2' ... c_n'} p(w_i \mid w_{i-2} w_{i-1}) \qquad (6)$$

where we borrow $\in$ to denote the "sub-string of" relation.

Another critical issue in sentence smoothing is that simply chaining up chunks in term of preferable ordering may not be enough to achieve an acceptable readability. Instead, some additional words (or chunks), such as function words, may be required for better readability. For example, when an English noun phrase is translated into Chinese, a classifier needs to be inserted. Thus, it is necessary to incorporate word insertion into the sentence generation model as given in (5), through taking into consideration a close set of smoothing words as possible chunks for generation — if adding some such chunk(s) can lead to high probability, we add them, and get a more readable sentence.

Other Issues

In addition to the four stages as discussed above, there are also many other important issues involved in EBMT, for example, user interface, example filtering and pattern inference, learning and usage modelling, etc., that we have not discussed here.

As far as the user interface is concerned, Kay's "translation amanuensis" position still holds: translation is a kind of editing work that needs a powerful interface to give all kinds of support to an editor, including retrieving all relevant translated fragments and memorizing new translation. Memorizing new translation is also a suitable human-machine interactive approach to example acquisition. Imagine what a powerful MT system we could finally produce if we had a distributed MT system for thousands of professional translators to use and input examples during their translation.

In addition to the input from its users, the machine's own learning ability beyond memorization is also important. Such learning tasks include learning collocation and fixed expressions from a given corpus. There are empirical collocation extraction techniques (e.g., Smadjia, 1993; Smadjia and McKeown, 1994) that we can employ.

The other learning tasks are to infer patterns or templates from existing examples and to do example filtering for example control. There is no doubt that such learning involves difficult problems, because we are obviously in a dilemma: we need to control the example number but within a case-based learning approach no individual example should be dropped in principle. Thus, the question to ask is: how do we know which examples are useful and which ones are not? Usage frequency information can be rather useful, but a measure for the usefulness of examples is to be defined in a more theoretically defendable way. Then, we can, hopefully, find a better strategy for example filtering.

Conclusion

In general, EBMT tackles the following problem: given the translation for the fragments of a source utterance, including its words, phrases and other nonconstituent chunks, infer the best choice of its translation in the target language with respect to the available translated fragments. In addition to a large-scale corpus in the target language, the translated fragments are the only resources for inference. There must be at least a sequence of fragments that covers the entire input utterance; otherwise, the input cannot be translated completely. The fragments may or may not be adequate in number for inferring a well-formed high-quality sentence in the target language, but we do want it to be as readable as possible. To enhance the capability of translation, it is necessary to collect translated fragments from existing parallel corpora, via text alignment and example acquisition discussed above.

In the above sections, we have given an overview on the EBMT approach to machine translation. We discussed the main issues involved, including its philosophy, origins, history and the four stages of work involved. We consider EBMT as an empirical case-based knowledge engineering approach to MT, in which the major means for knowledge acquisition is example acquisition by text alignment from large-scale parallel bilingual corpora. We reviewed the current state of the text alignment technology and recognized the importance of performing alignment at various linguistic levels for the acquisition of a comprehensive collection of examples. We also discussed some critical ideas about how to apply existing examples to translate new utterances, via source sentence decomposition and target sentence synthesis with the aid of language models.

The underlying principle for EBMT is as simple as this: remember everything translated in the past and use everything available to facilitate the translation of the next utterance. We know computers are the most fantastic machines to memorize such things as text pairs and their frequencies, and we thus have reason to believe that EBMT represents the machine translation approach with the greatest potential.

Acknowledgements

This work is part of the research output from the CERG project "EBMT for Hong Kong Legal Texts" funded by the Hong Kong University Grants Committee under the grant #9040482, with Jonathan J. Webster as the principal investigator and Kit Chunyu, Caesar Lun Suan, Pan Haihua, Sin King Kui and Vincent Wong as investigators. The authors wish to thank all team members who have contributed to the research work that enables this paper. Correspondence concerning this work should be addressed to Dr. Jonathan J. Webster, Department of Chinese, Translation and Linguistics, City University of Hong Kong, Tat Chee Ave., Kowloon, Hong Kong.

References

Arthern, Peter J. (1978). "Machine translation and Computerized Terminology Systems: A Translator's Viewpoint," in *Translating and the Computer: Proceedings of a Seminar*. London: Aslib, pp. 77–108.

BLIS (1998). *Bilingual Laws Information System* (BLIS). Hong Kong: Information Technology and Resources Unit, Administrative Division,

Department of Justice, Hong Kong SAR Government. Information available from http://www.justice.gov.hk/.

Brown, Peter F., John Cocke, Stephen A. Della Pietra, Vincent J. Della Pietra, Frederick Jelinek, J. D. Lafferty, Robert L. Mercer and Paul S. Roosin (1990). "A Statistical Approach to Machine Translation." *Computational Linguistics*, Vol. 16, pp. 79–85.

Brown, Peter F., Jennifer C. Lai and Robert L. Mercer (1991). "Aligning Sentences in Parallel Corpora," in *ACL-91*. Berkeley, 169–76.

Brown, Peter J., Stephen A. Della Pietra, Vincent J. Della Pietra, J. D. Lafferty and Robert L. Mercer (1993). "The Mathematics of Statistical Machine Translation: Parameter Estimation." *Computational Linguistics*, Vol. 19, pp. 263–311.

Brown, Ralf D. (1996). "Example-based Machine Translation in the Pangloss System," in *COLING-96*, pp. 169–74.

Chen, Stanley (1993). "Aligning Sentences in Bilingual Corpora Using Lexical Information," in *ACL-93*. Columbia, Ohio, pp. 9–16.

Church, Kenneth W. (1993). "Char-Align: A Program for Aligning Parallel Texts at the Character Level," in *ACL-93*. Columbia, Ohio, pp. 1–8.

Collins, B. and P. Cunningham (1995). "A Methodology for Example-based Machine Translation," in *CSNLP-95: Fourth Conference on the Cognitive Science of NLP Proceedings*. Dublin.

Dagan, Ido, Kenneth W. Church and William A. Gale (1993). "Robust Bilingual Word Alignment for Machine Aided Translation," in *Proceedings of the Workshop of Very Large Corpora*. Columbus, Ohio, pp. 1–8.

Evans, R. and A. Kilgarriff (1995). "MRDs, Standards and How to Do Lexical Engineering," in *Proceedings of Second Language Engineering Convention*. London, pp. 125–32.

Fung, Pascale and Kathleen McKeown (1997). "A Technical Word- and Term-translation Aid Using Noisy Parallel Corpora Across Language Groups." *Machine Translation*, Vol. 12, pp. 53–87.

Gale, William A. and Kenneth W. Church (1991). "A Program for Aligning Sentence in Bilingual Corpora," in *ACL-91*. Berkeley, pp. 177–84.

Hutchins, W. John (1998). "The Origins of the Translator's Workstation." *Machine Translation*, Vol. 13, pp. 287–307.

Kay, Martin (1997). "The Proper Place of Man and Machines in Language Translation." *Machine Translation*, Vol. 12, pp. 3–23. First printed as research report CSL-80-11, Xerox PARC, Palo Alto, California, 1980.

Kay, Martin and M. Roscheisen (1993). "Text Translation Alignment." *Computational Linguistics*, Vol. 19, pp. 75–102. First printed as Technical Report P90-00143 in Xerox Palo Alto Research Center in 1988.

Kit, Chunyu and Jonathan J. Webster (1992). "Machine Translation of Idioms Based on Tokenization," in *Proceedings of the First Singapore International Conference on Intelligent Systems*. Singapore.

Melby, Alan K. (1995). *The Possibility of Language: A Discussion of the Nature of Language*. Amsterdam and Philadelphia: John Benjamins Publishing Company.

Nagao, Makoto (1984). "A Framework of a Mechanical Translation Between Japanese and English by Analogy Principle," in A. Elithorn and R. Banerji, eds., *Artificial and Human Intelligence*. Amsterdam: North-Holland, pp. 173–80.

Nirenburg, Sergei, S. Beale and C. Domashnev (1994). "A Full-text Experiment in Example-based Machine Translation," in *International Conference on New Methods in Language Processing*. Manchester, pp. 78–87.

Nirenburg, Sergei, C. Domashnev and D. Grannes (1993). "Two Approaches to Matching in Example-based Translation," in *TMI'93*. Kyoto, pp. 47–57.

Pan, Haihua (1986). "A Machine Translation System for Scientific Titles of English," in *Proceedings of 1986 International Conference on Chinese Computing*. Singapore.

Pan, Haihua (1987). "Towards Understanding-based MT," in *Proceedings of the International Conference on Chinese Information Processing*. Beijing.

Pappegaaij, B. C., Victor Sadler and A. P. M. Witkam (1986). *Word Export Semantics: An Interlingual Knowledge-based Approach*. Dordrecht: Reidel.

Sato, Satoshi (1993). "Example-based Translation of Technical Terms," in *TMI'93*. Kyoto, pp. 58–63.

Sato, Satoshi and Makoto Nagao (1990). "Toward Memory Based Translation," in *COLING-90*. Helsinki, pp. 247–52.

Schubert, Klaus (1986). "Linguistic and Extra-linguistic Knowledge: A Catalogue of Language-related Rules and Their Computational Application in Machine Translation." *Computer and Translation*, Vol. 1, pp. 125–52.

Sin, King Kui and D. Roebuck (1996). "Language Engineering for Legal Transplantation: Conceptual Problems in Creating Common Law Chinese." *Language and Communication*, Vol. 16, pp. 235–54.

Smadjia, F. A. (1993). "Retrieving Collocation from Text: Xtract." *Computational Linguistics*, Vol. 19, pp. 143–77.

Smadjia, F. A. and Kathleen R. McKeown (1994). "Translating Collocation for Use in Bilingual Lexicon," in *Proceedings of the ARPA Human Language Technology Workshop*. Princeton.

Somers, Harold L. (1998). "New Paradigms in MT: The State of Play Now that the Dust Has Settled," in *Machine Translation Workshop, ESSLLI-98*. Saarbruecken, Germany, pp. 22–33.

Somers, Harold L. (2000a). "Example-based Machine Translation," in Russel Dale, H. Moisl and Harold L. Somers, eds., *Handbook of Natural Language Processing*. New York: Marcel Dekker, pp. 611–27.

Somers, Harold L. (2000b). "Machine Translation," in Russel Dale, H. Moisl and Harold L. Somers, eds., *Handbook of Natural Language Processing*. New York: Marcel Dekker, pp. 329–46.

Sumita, Eichiro, Osamu Furuse and Hiroshi Iida (1993). "An Example-based Disambiguation of Prepositional Phrase Attachment," in *TMI'93*. Kyoto, pp. 80–90.

Sumita, Eichiro, Hiroshi Iida and H. Khyama (1990). "Translating with Examples: A New Approach to Machine Translation," in *TMI'90*. Texas, pp. 203–12.

Webster, Jonathan J. and Kit Chunyu (1992). "Tokenization for Machine Translation: What Can Be Learned From Chinese Word Identification," in *Proceedings of 3rd International Conference on Chinese Information Processing*. Beijing.

Wilks, Yorick and M. Stevenson (1998). "The Grammar of Sense: Using Part-of-speech Tags as a First Step in Semantic Discrimination." *Natural Language Engineering*, Vol. 4, pp. 135–44.

Wu, Dekai (2000). "Alignment," in Russel Dale, H. Moisl and Harold L. Somers, eds., *Handbook of Natural Language Processing*. New York: Marcel Dekker, pp. 415–58.

Wu, Dekai (1995a). "Grammarless Extraction of Phrasal Translation Examples from Parallel Texts," in *TMI'95*. Leuven, Belgium, pp. 354–72.

Wu, Dekai (1997). "Stochastic Inversion Transduction Grammars and Bilingual Parsing of Parallel Corpora." *Computational Linguistics*, Vol. 23, pp. 377–404.

Wu, Dekai (1995b). "Stochastic Inversion Transduction Grammars, with Application to Segmentation, Bracketing, and Alignment of Parallel Corpora," in *IJCAI-95*. Montreal, pp. 1328–35.

Wu, Dekai and Xia X. (1995). "Large-scale Automatic Extraction of an English-Chinese Translation Lexicon." *Machine Translation*, Vol. 9, pp. 285–313.

Yarowsky, D. (2000). "Word-sense Disambiguation," in Russel Dale, H. Moisl and Harold L. Somers, eds., *Handbook of Natural Language Processing*. New York: Marcel Dekker, pp. 629–54.

The Application of Semantic Web Technology for Example-based Machine Translation (EBMT)

Jonathan J. Webster, Sin King Kui and Hu Qinan
Department of Chinese, Translation and Linguistics
City University of Hong Kong

Introduction

The Example-based Machine Translation Project at City University of Hong Kong applies the "example-based" approach to the translation of the specialised language of legislation and legal documents. Our purpose is twofold. First, we aim to meet the growing demand for bilingual legal texts as Hong Kong's legal system converts from a monolingual to a bilingual legal system. Second, we aim to explore the full potential of the example-based approach. Our research into the application of the example-based approach is based on an aligned parallel corpus representing the work of top professionals in legal translation. Initially our task is to design a best-match algorithm for translated text spans ranging in size and scope from words to phrases, clauses, and sentence patterns. The algorithm will be rigorously tested, and human input of improved translations will be constantly incorporated into the corpus in order to build up and develop the learning ability of the algorithm, thus enhancing the accuracy, consistency and intelligibility of the translated text.

The main issues in EBMT are example acquisition, example application and example base management. Focusing on these three aspects, we are pursuing a practical implementation of an EBMT system specialised for Hong Kong's bilingual English-Chinese legal document processing. Example acquisition in an EBMT system is concerned with how examples are acquired from previous translations. Simply put, an example is a pair of translated text spans. One approach to example acquisition is through text alignment on a parallel bilingual corpus at various linguistic levels, including word, phrase, clause and sentence. The 25 million-word bilingual English-Chinese legal document corpus is ideally suited for our purpose. The example application phase is concerned with how existing examples are used to facilitate translation. The main issues include: (a) identification of useful examples in an input sentence; (b) determination of a sequence (or chain) of identified examples to be used in composing the translation; and (c) further manipulation of the target language parts, e.g. reordering, to render the composition. Actually, this is the translation process. The third aspect concerns the management of the example base. Examples must be stored in such a way as to facilitate subsequent retrieval. Drawing on advances in semantic web technology, we employ the World Wide Web Consortium's recommended technology for metadata, i.e. RDF (Resource Description Framework), to tag and identify information about the corpus and its examples. The tags are a kind of vocabulary for describing the data, i.e. metadata. The vocabulary for describing the Bilingual Laws Information System (BLIS) corpus extends the basic *Dublin Core* metadata set with an enriched descriptive vocabulary for referring to aligned examples.

Bilingual Laws Information System (BLIS) Corpora

Source of the Database

The corpus contains the complete database of the statutory laws of Hong Kong stored in a CD-ROM provided by the technical unit of the Department of Justice. The general public may also access this database via the Bilingual Laws Information System (BLIS) on the Internet (http://www.justice.gov.hk). They may, however, only view or download the laws one section at a time. Given the enormous size of the database, corpus building would have been impossible without the full support of the Department of Justice of the Hong Kong SAR Government.

Size of the Corpora

The statutory laws of Hong Kong are divided into three main categories, namely, public ordinances (i.e., laws which concern the general public), private ordinances (i.e., laws which concern individual bodies, whether statutory or otherwise), and miscellaneous ordinances (i.e., laws which do not belong to either of the preceding categories). To date, there are 564 public ordinances, 166 private ordinances and 12 miscellaneous ordinances, adding up to a total of 742 ordinances (approximately 9 million English words and 5 million Chinese characters).

Structure of Hong Kong Ordinances

Hong Kong ordinances are arranged by chapters. Each chapter can be identified by its assigned number and short title, e.g., Cap 1, *The Interpretation and General Clauses Ordinance* (第一章 釋義及通則條例). Chapters 1–564 are the public ordinances and Chapters 1001–1166 the private ordinances. The numbers assigned to these two categories of ordinance are *official* numbers, i.e., they are the numbers appearing in the Loose-leaf Edition of the Laws of Hong Kong, which is the most authoritative version of the Laws of Hong Kong. The miscellaneous ordinances are also assigned numbers (from 2401, e.g., Cap 2401 *The National Flag and National Emblem Ordinance*) in the BLIS, but these are *unofficial* numbers, i.e., such numbers do not appear in the Loose-leaf Edition of the Laws of Hong Kong, but are assigned to the miscellaneous ordinances purely for the sake of compiling the database of the BLIS.

The content of an ordinance is, except for the Long Title, divided up and identified according to a very rigid numbering system.

a. *Parts*, identified by upper case Roman numerals, I, II, III, IV, … etc.
b. *Sections*, identified by Arabic numerals, 1, 2, 3, 4, … etc.
c. *Subsections*, identified by Arabic numerals in brackets (1), (2), (3), (4), … etc.
d. *Paragraphs*, identified by lower case letters in brackets (a), (b), (c), (d), … etc.
e. *Subparagraphs*, identified by lower case Roman numerals (i), (ii), (iii), (iv), … etc.

Parts and sections are also given headings, which are, however, not an operative part of an ordinance. Many ordinances contain schedules and/or forms identifiable by Arabic numerals.

The Chinese version of an ordinance follows the exact numbering system of its English counterpart. Chinese numerals are not used.

Accordingly, the Chinese texts of Hong Kong laws are perfectly aligned with the English texts in terms of chapters (*zhang* 章), parts (*bu* 部), sections (*tiao* 條), subsections (*kuan* 款), paragraphs (*duan* 段) and subparagraphs (*ji* 節). It is this feature that makes the bilingual texts of Hong Kong laws particularly suitable for the project because a well-aligned text of this size is seldom readily available.

Excerpts from the corpus are illustrated below:

CAP. 71 Control of Exemption Clauses PART II CONTROL OF EXEMPTION CLAUSES Avoidance of liability for negligence, breach of contract, etc. ... 8. Negligence liability (1) This section applies as between contracting parties where one of them deals as consumer or on the other's written standard terms of business. (2) As against that party, the other cannot by reference to any contract term– 　(a) when himself in breach of contract, exclude or restrict any liability of his in respect of the breach; or 　(b) claim to be entitled–	第 71 章 管制免責條款條例 第 II 部 管制免責條款 逃避因疏忽、違約等而引致的 法律責任 ...... 8. 合約因致的法律責任 (1) 如立約一方以消費者身分交易，或按另一方的書面標準業務條款交易，則本條適用於處理立約各方之間的問題。 (2) 對上述的立約一方，另一方不能藉合約條款而— 　(a) 在自己違反合約時，卸除或局限與違約有關的法律責任；或 　(b) 聲稱有權—

Semantic Web Technologies

Uche Ogbuji (2000) describes Extensible Markup Language (XML) as simply a data-format standard, which has made possible "more glamorous technologies" such as RDF. The Resource Description Framework (RDF) is the World Wide Web Consortium's recommendation for processing metadata. As Ogbuji explains, "RDF defines a directed graph of statements that describe Web-based resources." These statements about web resources are what are known as "metadata," i.e. data about data. RDF provides a framework for representing the semantics of the information being described. Moreover, by applying graph-traversal techniques to the structured metadata, data which had previously only been machine-readable is now rendered machine-processable.

Resource Description Framework

RDF (Lassila and Swick, 1999) represents metadata in terms of RDF statements, each of which is in the form of a triple composed of {*subject, predicate, object*}, where

- *subject* is the resource being described;
- *predicate* indicates which property the metadata is about;
- *object* is the value of that metadata.

As its name implies, RDF provides a common framework to describe kinds of metadata with respect to resources (Bray, 1998). The most familiar metadata are attribute-value pairs. For instance, the meaning of the sentence "John Smith is the author of the resource http://www.yahoo.com/," "john" will be expressed in an RDF statement as {http://www.yahoo.com/ john, author, "John Smith"}. Here *author* is one of the attributes of the resource "http://www.yahoo.com/john," and "John Smith" is the value of that attribute. In addition, RDF can also describe the metadata representing relationships among resources. Take as an example the sentence "The resource http://mail.yahoo.com is linked to the resource http://www.yahoo. com." Then the RDF statement will be {*http://mail.yahoo.com, be linked to, http://www.yahoo.com*}, denoting that the relationship between "http://mail.yahoo.com" and "http://www.yahoo.com" is *be linked to*.

Semantic Web

It will be easier to understand the concept of a semantic web if we compare it with the semantics of words. Research in lexical semantics has shown that the semantic relations among words, such as synonymy, meronymy, and hyponymy, help to capture word meaning in a systematic way. Originally, these relationships were kept separately within individual words. Making them explicit brings us a better understanding of each word. Furthermore, based on these relationships, words can be linked together to form semantic networks along which the search for word meaning can be performed. *WordNet* (Fellbaum, 1998), for example, has successfully established semantic hierarchies of large volumes of English words. Such relational information is much easier to represent and interpret by machine.

This architecture can be extended to the description of web resources. The meaning of web resources can be expressed to some extent by identifying their interrelationships and attributes. Based on this information, semantic hierarchies of resources, namely the semantic web (Lee, 1998; Brickley, 1999; Dumbill, 2000), can be generated. However, several problems exist in practice. For instance, the volume of the semantic information involved in a semantic web is huge. Thus, automating the processing of this information is critical.

As a general data model, RDF is employed as a key technology to assist with solving the above problem. The semantic web may help us to make the most of the information possessed by each resource. Since these resources are connected, inferences can be drawn by traversing through the nodes of resources according to any given queries. Thanks to the simplicity of the RDF data model, each RDF statement is automatically processable by computers. Up to the present, several RDF inference languages and engines have been developed to meet this demand (Guha *et al.*, 1998; Decker *et al.*, 1998).

Generally speaking, RDF enables the web to evolve from an opaque document repository to a rich knowledge base.

Extensible Markup Language

While RDF defines the basic conceptual model of the semantic web, Extensible Markup Language (XML) (Bray *et al.*, 1998) provides a user-friendly syntax to represent it. The RDF statement cited earlier can be serialized in XML as follows:

```
<rdf:RDF xmlns:rdf = "http://www.w3.org/1999/02/22-rdf-syntax-ns#">
<rdf:Description about = "http://www.yahoo.com/John">
<rdf:Author>John Smith</rdf:Author>
</rdf:Descripiton>
</rdf:RDF>
```

Although the XML serialization of RDF is referred to as RDF / XML, XML is only one possible syntax for RDF and alternate ways to represent the same RDF model may emerge.

Applications

Semantic web technology is being employed to turn the data from the BLIS corpora into machine-understandable information. We will illustrate how the corpora have been recoded using RDF/XML in order to combine the semantics of the metadata standard known as *Dublin Core* with a linguistically-rich vocabulary to facilitate subsequent linguistic processing and exploration.

Corpora in RDF / XML

Originally, the BLIS corpora were stored in *Lotus Notes*, which uses a proprietary database format. After the transformation from *Lotus Notes* into RDF/XML, the corpora are divided into two parts. The first part is the content of the parallel English/Chinese texts, which are to be aligned in terms of words, phrases and clauses. As illustrated in the following, each *<blis:section>* element contains one section (*tiao* 條) of the text, which is identified by the *<rdf:ID>* number.

```
<?xml version="1.0" encoding="UTF-8" ?>
<rdf:rdf  xmlns:rdf="http://www.w3.org/1999/02/22-rdf-syntax-ns#"  xmlns:
    dc="http://purl.org/dc/elements/1.1/"    xmlns:blis="http://cpct86.cityu.
    edu.hk/blis/">
…
<blis:section rdf:ID = "27811">PART VIA STATUTES The Council may
    make statutes for the administration of the University and for matters
    that this Ordinance provides for inclusion in a statute. (Added 92 of 1994
    s. 28)</blis:section>
<blis:section rdf:ID = "27812">第VIA部 規程 校董會可就大學的行政及本條
    例規定列入規程內的事宜訂立規程。（第VIA部由1994年第92號第28條增
    補）</blis:section>
…
</rdf:rdf>
```

Body Text in RDF/XML

The second part is the metadata, which describes the above resources. Each *<rdf:description>* element corresponds to one section of the above parallel texts.

```
<?xml version="1.0" encoding="UTF-8" ?>
<rdf:rdf  xmlns:rdf="http://www.w3.org/1999/02/22-rdf-syntax-ns#"  xmlns:
    dc="http://purl.org/dc/elements/1.1/"    xmlns:blis="http://cpct86.cityu.
    edu.hk/blis/">
…
<rdf:description about = "http://cpct86.cityu.edu.hk/blis/section#27811">
<dc:language>English</dc:language>
<dc:date>06/30/1997</dc:date>
```

```
<dc:identifier>http://cpct86.cityu.edu.hk/blis/section#27811</dc:identifier>
<blis:documentType>Ordinance</blis:documentType>
<blis:sectionType>Section </blis:sectionType>
<blis:chapterTitle> CITY UNIVERSITY OF HONG KONG </blis:chapterTitle>
<blis:legislationTitle> CITY UNIVERSITY OF HONG KONG </blis:
    legislationTitle>
<blis:sectionTitle> Statutes </blis:sectionTitle>
<blis:ordinanceIdentifier>1132</blis:ordinanceIdentifier>
<blis:sectionIdentifier>21A</blis:sectionIdentifier>
<blis:partIdentifier>I</blis:partIdentifier>
<blis:references />
<blis:definition />
<blis:remarks />
<blis:gazetteNumber />
<blis:alignment />
<blis:source />
</rdf:description>
<rdf:description about = "http://cpct86.cityu.edu.hk/blis/section#27812">
<dc:language>Chinese</dc:language>
<dc:date>06/30/1997</dc:date>
<dc:identifier>http://cpct86.cityu.edu.hk/blis/section#27812</dc:identifier>
<blis:documentType>Ordinance</blis:documentType>
<blis:sectionType>Section</blis:sectionType>
<blis:chapterTitle>香港城市大學條例</blis:chapterTitle>
<blis:legislationTitle>香港城市大學條例</blis:legislationTitle>
<blis:sectionTitle>規程</blis:sectionTitle>
<blis:ordinanceIdentifier>1132</blis:ordinanceIdentifier>
<blis:sectionIdentifier>21A</blis:sectionIdentifier>
<blis:partIdentifier>I</blis:partIdentifier>
<blis:references />
<blis:definition />
<blis:remarks />
<blis:gazetteNumber />
<blis:alignment />
<blis:source />
</rdf:description>
...
</rdf:rdf>
```

Metadata in RDF/XML

The discrimination of the parallel texts and the metadata could facilitate subsequent alignment process significantly. Besides being based on the translation pairs in the dictionary, our alignment algorithm relies heavily on the metadata of each section. By comparing the values of *ordinanceIdentifier, sectionIdentifier* and *partIdentifier*, the location of the aligned texts can be easily identified.

Dublin Core

It is clear that much of the metadata shares a common semantics with other applications, such as *language*, *date* and *source*. To maintain consistency with other applications, we use the *Dublin Core* Metadata Element Set (Dublin Core Working Group, 1999).

Dublin Core was originally developed to improve resource discovery on the web. It is composed of the following fifteen elements, which are

Dublin Core Metadata Element Set

Metadata	Meaning	Category
coverage	The extent or scope of the content of the resource.	Content
description	An account of the content of the resource.	
type	The nature or genre of the content of the resource.	
relation	A reference to a related resource.	
source	A Reference to a resource from which the present resource is derived.	
subject	The topic of the content of the resource.	
title	A name given to the resource.	
contributor	An entity responsible for making contributions to the content of the resource.	Intellectual Property
creator	An entity primarily responsible for making the content of the resource.	
publisher	An entity responsible for making the resource available.	
rights	Information about rights held in and over the resource.	
date	A date associated with an event in the life cycle of the resource.	Instantiation
format	The physical or digital manifestation of the resource.	
identifier	An unambiguous reference to the resource within a given context.	
language	A language of the intellectual content of the resource.	

generally assumed to be widely applicable to the description of web resources.

Moreover, *Dublin Core* provides a mechanism to further refine the meaning of above elements for domain specific needs. In our application, we designed our own RDF Schema (Brickley and Guha, 2000) to define qualifiers (Dublin Core Working Group, 2000; Miller, Miller and Brickley, 1997) for interpretation of the elements.

An RDF Schema specifies the semantics of the statements given in an RDF data model, that is the semantics of the attributes and relationships and constraints between resources. The RDF Schema core vocabulary can be used to make RDF statements defining application-specific vocabularies. The element qualification of *Dublin Core* in BLIS Schema is shown below:

BLIS Qualifiers

Dublin Core	BLIS Qualifiers
identifier	ordinanceIdentifier, sectionIdentifier, partIdentifier
title	chapterTitle, legislationTitle, sectionTitle
type	documentType, sectionType
relation	references, alignment

Alignment Results in RDF/XML

The clausal and phrasal alignment results generated in our EBMT project could also be represented in RDF/XML format.

```
<?xml version="1.0" encoding="UTF-8" ?>
<rdf:rdf  xmlns:rdf="http://www.w3.org/1999/02/22-rdf-syntax-ns#"  xmlns:
    dc="http://purl.org/dc/elements/1.1/"  xmlns:blis="http://cpct86.cityu.edu.
    hk/blis/">

...

< blis:phrase rdf:ID = "14789"> for matters that this Ordinance provides for
    inclusion in a statute </rdf:phrase>
< blis:phrase rdf:ID = "14790"> may make statutes </rdf:phrase>

...

< blis:phrase rdf:ID = "15213"> 本條例規定列入規程內的事宜</rdf:phrase>
< blis:phrase rdf:ID = "15214"> 訂立規程</rdf:phrase>

...

</rdf:rdf>
```

Alignment Results in RDF / XML

```
<rdf:description about = "http://cpct86.cityu.edu.hk/blis/phrase#14789">
...
<dc:identifier>http://cpct86.cityu.edu.hk/blis/phrase#14789</dc:identifier>
<blis:alignment>http://cpct86.cityu.edu.hk/blis/phrase#15213</blis:alignment>
<blis:source>http://cpct86.cityu.edu.hk/blis/clause#12287</blis:source>
...
</rdf:description>
```

Metadata in RDF / XML

The alignment information is kept in the metadata. The above metadata shows that the source of the phrase#14789 is clause#12287, and the phrase#14789 and phrase#14790 are aligned phrases. By this means, the whole corpora are linked together.

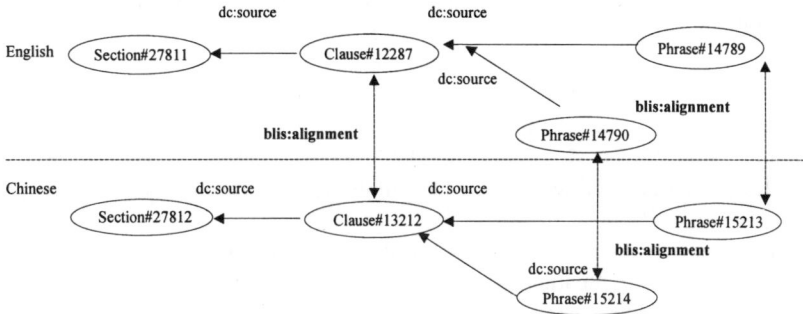

Not only are the aligned segments maintained, the relationships between segments and the corpora are also realized. It would be easy to trace a piece of English phrase back to its corresponding original Chinese section.

Conclusion

By applying semantic web technologies, the BLIS corpus, which is originally a flat database, has been developed to be a solid machine-understandable knowledge base.

Being machine-understandable information, the metadata can unify and automate a large number of information management tasks upon large volumes of data, including facilitating the alignment process in the example acquisition phase of EBMT. It is predictable that the metadata will also help the example base management tasks. As for example application, the knowledge base has the potential to be a well-established exploration space for more efficient subtree-matching.

References

Bray, Tim (1998). "RDF and Metadata." http://www.xml.com/pub/98/06/rdf.html, 9 June.

Bray, Tim, Jean Paoli and C. M. Sperberg McQueen (1998). "Extensible Markup Language (XML) 1.0." http://www.w3.org/TR/1998/REC-xml-19980210/, 10 February.

Brickley, Dan (1999). "Nodes and Arcs 1989-1999: The WWW Proposal and RDF: Then and Now." http://www.w3.org/1999/11/11-WWWProposal/thenandnow, 12 November.

Brickley, Dan and R. V. Guha (2000). "Resource Description Framework (RDF) Schema Specification 1.0" http://www.w3.org/TR/2000/CR-rdf-schema-20000327/, 27 March.

Decker, Stefan, Dan Brickley, Janne Saarela and Angele Jürgen (1998). "A Query and Inference Service for RDF." http://www.w3.org/TandS/QL/QL98/pp/queryservice.html,. 18 November.

Dublin Core Working Group (1999). "Dublin Core Metadata Element Set, Version 1.1: Reference Description." http://dublincore.org/documents/1999/07/02/dces/, 2 July.

Dublin Core Working Group (2000). "Dublin Core Qualifiers." http://dublincore.org/documents/1998/09/dces/, 11 July.

Dumbill, Edd. (2000). "The Semantic Web: A Primer." http://www.xml.com/pub/a/2000/11/01/semanticweb/index.html, 1 November.

Fellbaum, Christiane (1998). *WordNet: An Electronic Lexical Database.* Cambridge, Massachusetts: Massachusetts Institute of Technology Press.

Guha, Ramanathan V., Ora Lassila, Eric Miller and Dan Brickley (1998). "Enabling Inferencing." http://www.w3.org/TandS/QL/QL98/pp/enabling.html.

Lassila Ora and Ralph R. Swick (1999). "Resource Description Framework (RDF) Model and Syntax Specification." http://www.w3.org/TR/REC-rdf-syntax/, 22 February.

Lee, Tim Berners (1998). "Semantic Web Road Map." http://www.w3.org/DesignIssues/Semantic.html, 14 October.

Miller, Eric, Paul Miller and Dan Brickley (1997). "Guidance on Expressing the *Dublin Core* within the Resource Description Framework *(RDF)*." http://www.ukoln.ac.uk/metadata/resources/dc/datamodel/WD-dc-rdf/, 1 July.

Ogbuji, Uche (2000). "An Introduction to RDF: Exploring the Standard for Web-based Metadata." http://www-106.ibm.com/developerworks/library/w-rdf/, November.

PART 2

Terminology

Chinese IT Terminology Management in Hong Kong

Aman Chiu and Björn Jernudd
Department of English Language and Literature
Hong Kong Baptist University

Introduction

Terminology work, particularly the harmonisation and standardisation of terminology, has been carried out by subject specialists of all fields for centuries to facilitate efficient subject communication and knowledge organisation.Wherever and whenever specialised information and specialised knowledge are being prepared, processed, represented, transformed and transferred, terminology is accorded a crucial role. Wright and Budin (1997:327) refer to terminology management as "any deliberate manipulation of terminological information."

However, the literature on terminology processing reveals much about the principles of adjusting in general but does not discuss the sources and circumstances of problems. Also, while a fair amount is known about terminology management in Europe and the U.S., relatively little has been written about it in Hong Kong. There are considerable gaps in knowledge in this aspect, both with reference to Hong Kong's situation in general and subject areas in particular. To fill the gaps there are a series of questions which require answers. In terms of Chinese IT terminology, for example, we need to know:

(1) What are the terminology problems in Hong Kong?
(2) Why are there such problems?
(3) Who claims what to be problems?

(4) Who standardises the Chinese IT terminology in Hong Kong?
(5) Why and how do they standardise terminology?
(6) What if any are the actual discourse problems in using the terminology?
(7) How do people in discourse, online and offline, manage terminology?

The reason for this neglect is understandable: subject specialists who are capable of identifying and distinguishing the subtlety of technical concepts (such as IT) are not usually interested or linguistically trained to undertake this work; linguists, on the other hand, rarely have the detailed subject knowledge to produce reliable evidence for determining the regularities underlying processes of term formation and adjustment.

Terminology work in Hong Kong seems to be accorded little importance. Where there is no perceived need for terminology development, there are no funds for this work, and where there are no funds there can be no activities. In such cases it is only possible to develop theoretical positions. However, in the absence of testing grounds, these cannot then be developed into practical guidelines and methodologies. The type of terminology work done for many years in Europe, e.g., by TNC (a Swedish terminological centre) in Sweden and TSK (its Finnish equivalent) in Finland, is yet an uncharted domain of enquiry in Hong Kong; it is, therefore, not surprising that little progress has been made in developing standardised territory-wide applicable guidelines for the processing of technical terminology including those in IT areas.

Terminology work has always been described as an interdisciplinary activity (Sager,1990:3; Cabré, 1999:9) rather than a subject in its own right. To address oneself to so large a subject in the context of globalisation (one to which the next section is dedicated) is to invite trouble. As the subject could be approached in so many different ways, it is important at the outset to set the boundary for the analysis used in this paper.

This paper aims at explaining links between kinds of Chinese IT terminology problems and kinds of decision-making. The paper relies on the language management model. (Jernudd and Neustupný, 1987, 1991) This paper also aims at describing the Chinese IT terminology work carried out by standardisation bodies in Hong Kong. It highlights the significance of terminology unification in achieving efficient communication in such an emerging revolutionary communications environment as that of Hong Kong today.

Background to the Study

Cabré (1999:3) has noted a paradoxical phenomenon in terms of the status of languages in the context of globalisation:

> On the one hand there is a trend towards monolingualism across cultures which is justified by the need for direct and efficient communication; on the other hand, national languages are being recognized as the natural tools of communication at all levels of usage, whether general or specialized.

The rapid development of IT as well as worldwide networking of databases has made parallel usage of several languages a matter of course. The very basic need for people to communicate demands a local language. Talk amongst the technological elite in English is a discourse which has little meaning for a lot of people in many contexts of the society:

> I am convinced that exogenous languages cannot withstand the force of indigenous languages. Indigenous languages may be oppressed or their populations marginalized for whatever social, economic or political reasons to such extents that they perish. But as long as indigenous languages have support in a community of speakers, their very presence and use exert pressure of acceptance and of being given a place and of extension of their use from within families into broader social and public contexts. (Jernudd, 1997: 18)

Like all kinds of businesses, IT of various sorts is expanding its influence across national boundaries into culturally and linguistically foreign territories. Localisation follows in the wake of globalisation. Most people would prefer to communicate in their own language. This is the same as the preference for people to buy products made in their own country, that is, products that have interfaces and documentation in their usual languages.

It might be true that talk amongst the technological and educational elites is in most cases conducted in English, which is the *lingua franca* in the domain of sciences and technology. Such discourse, however, has little meaning for a lot of people in many contexts of society. Wang (2000:38) highlights the significance of translation (that is, localisation) in the present stage of globalisation since translation always stands between two languages and societies and thus it assumes an important role of "intercultural communication in a context of globalisation."

In Hong Kong, localisation of IT terminology in a lot of cases surfaces as ad-hoc bilingual glossaries in newspapers or magazines. It also surfaces as formal glossaries by professional or governmental bodies. Since 1995, for example, the Hong Kong Computer Society has systematically been trying to further the use of standardised Chinese IT terms to promote effective communications within the trade. Three government bodies, namely the Information Technology Services Department, the Official Languages Agency, and the Curriculum Development Institute Council, have also since 1997 developed standardised IT bilingual glossaries for both internal and external communication.

Is There a Terminological Problem?

Laissez-faire?

In Hong Kong today, almost everyone comes in contact with computers. We are increasingly adopting new communications environment such as the Internet, e-mail and video conferencing. In most cases, when it comes to what could be agreed as the Chinese term for a specific IT term, there is variation which requires choice by translators, editors, interpreters, technical writers, and terminologists. How should they choose? And if there is a gap, how should they proceed? The word "Internet," for instance, has been rendered into at least four different translations (data taken from *Chinese Database Development Bulletin* 《詞庫建設通訊》, No. 17, September 1998, published by the Hong Kong Chinese Language Society 香港中文語文學會):

Translations	Pronunciation*	Literal meanings
1. 互聯網	*wuh lyuhn mohng* *hu lian wang*	互 mutually / 聯 united / 網 nets
2. 互連網	*wuh lihn mohng* *hu lian wang*	互 mutually / 連 linking up/ 網 nets
3. 網際網	*mohng jai mohng* *wang ji wang*	網 net / 際 inter-; among / 網 nets
4. 網間網	*mohng gaan mohng* *wang jian wang*	網 net / 間 among; space in between / 網 nets

* top line denotes Cantonese pronunciation; bottom line denotes Putonghua (same for the next table)

The term "Internet" has even more Chinese names (*ibid.*):

Translations	Pronunciation	Literal meanings
1. 互聯網	*wuh lyuhn mohng* *hu lian wang*	互 mutually / 聯 united / 網 nets
2. 國際網絡	*gwok jai mohng lohk* *guo ji wang luo*	國際 international / 網絡 network
3. 國際資訊網絡	*gwok jai ji seun mohng lohk* *guo ji zi xun wang luo*	國際 international / 資訊 information / 網絡 network
4. 國際互聯網	*gwok jai wuh lyuhn mohng* *guo ji hu lian wang*	國際 international / 互 mutually / 聯 united / 網 nets
5. 英特網	*ying dahk mohng* *ying te wang*	英特 *ying te*, transliteration of "*Inter*(net)" / 網 nets
6. 因特網	*yan dahk mohng* *yin te wang*	因特 *ying te*, transliteration of "*Inter*(net)" / 網 nets
7. 交互網	*gaau wuh mohng* *jiao hu wang*	交 cross; intersect / 互 mutually / 網 nets
8. 全球互連網	*chyuhn kauh wuh lyuhn mohng* *quan qiu hu lian wang*	全球 global / 互 mutually / 聯 united / 網 nets
9. 國際電腦網絡	*gwok jai dihn nouh mohng lohk* *guo ji dian lao wang luo*	國際 international / 電腦 computer / 網絡 network
10. 國際計算機 互聯網	*gwok jai gai syun gei wuh* *lyuhn mohng* *guo ji ji suan ji hu lian wang*	國際 international / 計算機 computer / 互 mutually / 聯 united / 網 nets

Many English IT terms have been rendered into various different Chinese translations. Some language authorities claim this is a problem — something which we will discuss below. Examples abound: *Java* can be *jia wa yu yan* 佳娃語言, *gua wa yu yan* 爪哇語言 or *Java-yu yan* Java 語言; a *mailing list* becomes *han jian fa song qing dan* 函件發送清單, *you jian qing dan* 郵件清單, *you jian lie biao* 郵件列表 or *you di lie biao* 郵遞列表; *virtual reality* has more than ten Chinese names, such as *dian xiang* 電象, *xu ni shi jng* 虛擬實景, *xu ni xian shi* 虛擬現實, *xu ni zhen shi* 虛擬真實, *xu jing* 虛境, *ni zhen* 擬真, *ling jing* 靈境 and *you zhen xian shi* 猶真現實. *Cyberspace* has even more, and they include: *wang luo kong jian* 網絡空間, *dian nao kong jian* 電腦空間, *dian xiang kong jian* 電像空間, *duo wei kong jian* 多維空間, *duo wei zi xun kong jian* 多維資訊空間, *yi du kong jian* 異度空間, *wang luo zi xun kong jian* 網絡資訊空間, *dian zi kong jian* 電子空間, *sai bo kong jian* 賽博空間, *shu zi shi kong* 數字時空, *dian nao shi jie* 電腦世界, *wang luo shi jie* 網絡世界, and so on.[1]

The Problem: Can We Tolerate Ambiguity?

How do people handle this rich offering of vocabulary? Comments in the *Newsletter of East Asia Forum on Terminology* (1999, 3:1) highlight the difficulties in standardisation work when faced with "bad" translations:

> "Bad" terms based on inappropriate translation, once established by popular usage, become hard to revise. We have to accept many of these bitter fruits while we meet with great resistance in popularising the corrected version. Moreover, few term systems have been established that conform with the standards set out in GB terminological documents. Lots of inconsistency between the terminology of various fields of science and technology remain to be harmonised.

When communicating parties attribute different meanings to the same word, or when there is insufficient communication because of choice of imprecise expression or different or inappropriate translation, the consequences can be costly or dangerous. Sometimes it costs only a loss of face but in a technical context it could cost a lot.

The problems may be compounded when the Chinese names are expressed differently in different Chinese-speaking regions, such as Hong Kong, Taiwan, China, Singapore, Malaysia, and so on. Ways of translating foreign terms into Chinese may be many. Translators and technical writers might wonder which name is more appropriate: 網絡空間 (*wang luo kong jian*/network space), 電腦空間 (*dian nao kong jian*/computer space), 電像空間 (*dian xiang kong jian*/electronic space), 多維空間 (*duo wei kong jian*/multi-dimensional space), etc., are all used to refer to *cyberspace*. Journalists and the public may not understand each other when 黑客 (*hei ke*/black guest) and 駭客 (*hai ke*/shocking, horrifying guest) are used for *hacker*; or when 電子信函 (*dian zi xin han*/electronic letter), 電子信件 (*dian zi xin han jian*/electronic mail), 電子函件 (*dian zi xin han jian*/electronic correspondence), or 電子信箱 (*dian zi xin xiang*/electronic post box) are used interchangeably for *e-mail*, even in the same language texts. News from 環球資訊網 (*huan qiu zi xun wang*/globalised information network) may not be seen as identical to that from 環球網 (*huan qiu wang*/globalised network), 萬維網 (*wan wei wang*/multiple dimensional network), or 全球瀏覽系統 (*quan qiu liu lang xi tong*/worldwide browsing system), all of these in fact refer to the WWW.

What problems do people claim there are? Who make such claims? Why? What are the actual discourse problems? Do claims and experience of problems coincide?

Who Claims What Problems?

This section examines how individual users such as terminologists, translators and subject specialists respond to terminology problems and what they claim to be the term problems.

Data Collection

The China National Committee for Terms in Sciences and Technologies (CNCTST) is authorised by the State Council of the PRC Government to examine and approve, promulgate and supervise the scientific and technological terms on behalf of the Chinese government. CNCTST is based in Beijing and is the principal standardisation body for terms in sciences and technologies in China. It also aims at comparing and analysing terminology used in Hong Kong and Taiwan in the hope of unifying usage.[2]

This section is based on an analysis of questions and discussions extracted from journals[3] published in both China and Hong Kong. These journals discuss terminology issues from different perspectives by users (contributors) including subject experts, information specialists, terminologists and translators. There is data on "who asks what questions and on why," and "who claims what are the problems," and the data often includes notes on suggested adjustment. The articles reveal how subject specialists, linguists or translators evaluate Chinese IT terms. They reveal aspects of meta-language, offer evidence of peoples' theories, if any, about vocabulary, and reveal both linguistic and non-linguistic interests.

The Term Conflicts

We organise our discussion according to a classification of term conflicts that characterise sets of term problems:

(1) Conflicts between a norm and its variants
(2) Conflicts between official norm and common usage
(3) Conflicts between two different norms carried by two populations
(4) Conflicts in the norm itself

Conflicts Between a Norm and Its Variants

Disk and Disc

A major debate flared up around giving an appropriate name for *disk* and

disc. It was apparently initiated by journal Issue No. 1, March 2000 of
CNCTST, and was followed by a number of articles contributed by four
subject specialists.[4]

Compact Disc and Its Variants

The standard term announced by the IT Services Department in Hong
Kong for *compact disc* is 光碟 *guang die* (small bright plate), but there are
many variants in actual use. Debaters charge there is confusion in usage
because of the availability of variants. One line of thinking (Zhang, Yang
and Lin) proposes that *disk* should be translated into 碟 *die* (small plate)
and *disc* into 盤 *pan* (plate), as *disk* and *disc* denote different concepts in
English. Another line of thinking (Tian) proposes that *disk* and *disc* should
be rendered with one term as the single translation.

Zhang observes that there are more than ten competing names for the
term CD (compact disc): CD 碟 (*CD die*/CD small plate), CD 盤 (*CD pan*/
CD plate), 光碟 (*guang die*/small bright plate), 壓印光碟 (*ya yin guang
die*/small pressed printed bright plate), and 壓縮光盤 (*ya suo guang die*/
small compressed bright plate), etc. He diagnoses this as a term problem,
since these terms deviate not only from the standard, but also from the
original concepts denoted by the term *disc* and *disk*. Zhang evaluates the
variation in usage as inadequate, and, therefore, a term problem, especially
in the contexts of "science and technology, manufacturing, commercial
communication and daily life."

Zhang describes the source of the problem as:

> *Disk* and *disc* were two same words with the same meaning in the past in
> China and in overseas countries, especially before the widespread use of disc.
> There was no distinction between the two terms and so they were used
> interchangeably. Since the middle of 1990s, however, *disk* and *disc* have been
> distinguished as separate terms by some influential and authoritative
> dictionaries of computing and related subjects published in the US and the UK
> ...

He quotes definitions from references to support his evaluation that
disc is "a round, flat piece of non-magnetic, shiny metal encased in a
plastic coating, designed to be read from and written to by optical (laser)
technology"; while *disk* is a "round, flat piece of flexible plastic or
inflexible metal coated with a magnetic material."

He evaluates that *disc* and *disk* refer to two different concepts, and
anticipates problems in discourse when the two terms are translated into

one single Chinese name. The one-to-one correspondence between the terminological and conceptual systems must not be violated lest there be misunderstanding and communication breakdown.

Zhang hence proposes a solution to the problem:

> *Disk* is used to signify *ci pan* 磁盤 (magnetic plate) and its derived compounds, such as magnetic disk, cartridge disk, floppy disk, hard disk, and disk array; and *disc* is used specifically for *guang die* 光碟 (small bright plate) and its derivatives, including optical disc, laser disc (LD), compact disc (CD), digital video disc (VCD), and disc array.

The adjustment he recommends is to render *disc* into *guang die* 光碟 (small bright plate), or *die* 碟 (small plate) in short; and *disk* into *ci pan* 磁盤 (magnetic plate), or *pan* 盤 (plate) in short, so as to distinguish the differences of *disc* from *disk* in the conceptual systems to avoid misunderstanding in discourse.

Fang Jing, the deputy head of the editorial office of CNCTST's journal, in one of her letters to the researcher, recommends:

> The terms *guang pan* 光盤 (bright plate) and *guang die* 光碟 (small bright plate) are problematic enough. We are prepared to adopt the views as proposed by Lin and Yang in Issue No. 1, 1998 and to collocate *guang* 光 (bright) with *die* 碟 (small plate), and *ci* 磁 (magnetic) with *pan* 盤 (plate). *Ci pan* 磁盤 (magnetic plate) will become the equivalent for *disk*, and *guang die* 光碟 (small bright plate) for *disc*. There are still different views here and we will later call upon a final meeting in order to standardize the names for them. (Fang Jing, letter to Aman Chiu, 30 June 2000)

Conflicts Between Official Norm and Common Usage

Another kind of term conflict concerns the enforcement of an official norm over a common usage. A case in point is the negotiation for a more appropriate name for *Internet*.

Internet: Hu Lian Wang vs Yin Te Wang

The term *hu lian wang* 互聯網 (mutually united nets) has been used widely to refer to the English term *Internet*. *Hu lian wang* 互聯網 (mutually united nets) is also a standard name used in Hong Kong. CNCTST in Beijing unanimously placed itself in 1997 behind using *yin te wang* 因特網 ("yin te" nets), which is the transliteration for *Internet*, but some users note it as a term problem, and want the simple *hu lian wang* 互聯網 (mutually united nets) to remain.

Fang (1999) criticizes the prescriptive function of CNCTST when *Internet* officially became *yin te wang* 因特網 ("yin te" nets). He also pinpoints a most likely cause for the continued use of *hu lian wang* 互聯網 (mutually united nets) to refer to *Internet* in discourse, especially in daily life:

> I am in school every day, and in here you almost cannot hear the word *yin te wang*, since *hu lian wang* has already become part of our school life here. Every time when I return to the town or the village, I can only hear *hu lian wang*, and not *yin te wang*. Let's imagine how difficult it is for people to understand you when you utter the suave word *yin te wang* to a worker or a farmer? Are you making fun of their knowledge?

Even when *Internet* was officially named *yin te wang* 因特網 ("yin te" nets), however, it was still referred to, in common usage, as *hu lian wang* 互聯網 (mutually united nets) according to Fang. As long as people communicate with each other, *hu lian wang* 互聯網 (mutually united nets) serves as a name in common usage to make sure the term refers to the English term *Internet* and thus the concept denoted by it. This is particularly so as the memory of referring *hu lian wang* 互聯網 (mutually united nets) to *Internet* still remains in people's minds. Communication may simply break down, Fang claims, if people choose to use the "suave word" *yin te wang* officially announced by the government to refer to *Internet*.

Confusion supposedly motivated this claimed term conflict. The simplest adjustment at the moment of discourse (e.g., with a worker) is simply to ask "which *net*" (*Hu lian* NET vs *Yin te* NET) one is referring to? In the overt management debate, the equivalent sequence of consideration could be described as follows: the expression *yin te wang* 因特網 ("yin te" nets) is noted as a deviation from the common usage *hu lian wang* 互聯網 (mutually united nets), and is therefore evaluated as inadequate and the adjustment to the common usage *hu lian wang* 互聯網 (mutually united nets) is suggested which in turn could be implemented as an agreed name which enhances efficient communication. On this ground Fang criticizes the misjudgment shown by CNCTST in suggesting the renaming of *Internet* into a name which is unfamiliar in common usage.

Conflicts Between Two Different Norms Carried by Two Populations

Term Problems in Cross-regional Communication

Another case to be considered is the existence of overt norms specific to markets. For instance, the norm in China or Taiwan may not be exactly the same as the norm in Hong Kong. Would term conflicts arise from the use of regionalisms or geographically restricted terms? Terminology management may encompass cases where one single form has to be chosen among two or more variants which are also accepted usage but in a different region or territory. In Hong Kong, one could say that certain IT terms are standardised in the sense that the civil service is required to use only one specific "official version" as listed in a government issued glossary, even if it runs against the rules of other official Chinese languages in mainland China or Taiwan. Thus, the term *Internet* is called *hu lian wang* 互聯網 (mutually united nets) in Hong Kong as standardised by the Official Languages Agency and the IT Services Department, although the standardisation body in China would dictate another name *yin te wang* 因特網 ("yin te" nets) and in Taiwan *guo ji wang ji wang luo* 國際網際網絡 (international inter-network).

The Adjustment of Switching to a "Variant Norm"

In contact discourse, deviations from the norm of one region are noted by certain users who are well aware of the differences in norms in different places. The choice of terms of one variety over another in the course of communication, for instance, is a technique to avoid misunderstandings that may occur from the use of words which are perceived differently in the varieties in Hong Kong and China. In such a case, a switching into what we should call a "variant norm" depending on who the audience are becomes a strategy that is implemented discoursally to facilitate communication. In China, for example, the two terms *internet* and *Internet* have different Chinese names as they refer to different concepts. The standard name for *Internet* is *yin te wang* 因特網 ("yin te" nets) and the one for *internet* is *hu lian wang* 互聯網 (mutually united nets). In Hong Kong, however, no differentiation is being made to distinguish *internet* from *Internet*, and the name *hu lian wang* 互聯網 (mutually united nets) is used to refer to both. Another example is the term *information technology*, the standardised name for which in use in China is *xin xi ji shu* 信息技術 (information/data technology), but *zi xun ke ji* 資訊科技 (information science and technology) in Hong Kong.

A norm conflict is noted as a problem, and a repair strategy of switching is selected as adjustment. This can be illustrated by discourse reproduced from the exchanges of faxes between the researcher and CNCTST.

In one of her letters to the co-author (Aman Chiu), Fang Jing at CNCTST notes a deviation of the term *zi xun ke ji* 資訊科技 (information science and technology) made by the researcher in his previous letter against the overt norm, which is *xin xi ji shu* 信息技術 (information/data technology) used in China. This constitutes a term conflict and thus inadequacy is established.

The correction adjustments Fang adopts are to switch to the variant that is adopted by the reader/hearer of the language, although she as a terminologist also makes a remark about the conceptual differences (thus another term problem noted) between the two terms. She said:

> … Concerning your enquiry about a English-Chinese database on *zi xun ke ji* 資訊科技 (the conceptual equivalence between *zi xun ke ji* 資訊科技 and *xin xi ji shu* 信息技術, the term used by us, is another topic which calls for discussion)… (⋯⋯關於您想要的英漢資訊科技術語詞庫的事 (『資訊』與我們叫的『信息技術』概念對應問題也是應研討的問題) ⋯⋯)

In communicating with the co-author, Fang chooses to switch to the norm used in Hong Kong (by the audience she is communicating with), although she also lists the norm used in China in brackets. Standardisation of Chinese technical terms is introduced in different regions, and there is a fundamental regional conflict between the need for naming and the desire to unify names. In his reply to CNCTST, Chiu manages his language in the same manner:

> … I have been conducting a research on *zi xun ke ji* 資訊科技 (information science and technology) (in your term "*xin xi ji shu* 信息技術 (information/data technology)") terminology after I quit my job as a lexicographer… (我辭掉了詞典編輯工作以後，就集中做這項資訊科技 (您們叫『信息技術』) 名詞研究工作⋯⋯)

> … Concerning your question about the translation of *disc* and *disk*, I have posted the question onto *hu lian wang* 互聯網 (the Internet) (in your term "*yin ta wang* 因特網") for discussion… (關於您提到的 *disc* 和 *disk* 的翻譯問題，我就此已把這問題放在互聯網 (您們叫『因特網』) 上討論⋯⋯)

Another example can be cited (Wang, 1999:5): the editor of the CNCTST journal switches to the norm expected by the audience of the journal by adding a note to the term made by a contributor from Singapore. The adjustment of switching is a problem-solution technique in anticipation of the incomprehensibility of a foreign norm, that is *wang ji wang luo* 網際網絡 (inter-nets network) used in Singapore for the term *Internet*, to the audience in China.

One can imagine that in discourse contact situations, with different norms being adopted by the two communicating parties, there are instances when one has to make certain adjustments (switching to a variant norm in the above cases) in one's writing (or speech production), either for the reason of a production or reception problem, the negotiation of meanings or in order to achieve certain intentions, that is, successful communication.

Conflicts in the Norm Itself

Inadequacy of the Existing Norm: Cyberport

Terminology management also encompasses cases where the standard status (norm) that consensus terms acquire is challenged, because of their being evaluated as inadequate in meaning. In such a case, a semantic interest directly motivates the adjustment to recommend a new norm to replace the existing one. One of the most convincing illustrations of this phenomenon is the Chinese term *shu ma gang* 數碼港 (digital port), a translation adopted by the Hong Kong government to refer to the equally new coinage *Cyberport*.

The Chinese rendering is perceived as inadequate by Wu (1999). (Wu was a former translator in the United Nations and is currently editor of an electronic publication.) *Shu ma gang* 數碼港 (digital port) was reanalysed as literally meaning "digital port," and so does not convey the complete sense of *Cyberport*, which should cover two layers of meanings, including not only the "digital" but also the "analog" perspectives. Wu analyses the term as:

> The term *cyber* in English has the meanings of both telecommunications and control, the means to which include modern telecommunications technology, i.e. digital and analog. 數碼 *shu ma* only covers the digital side of the meaning. (英語裡 *cyber* 一詞有電子通信和控制的意義，其手段則包括 *digital*〔數字，數碼〕和 *analog*〔模擬〕等現代電信技術。數碼僅包括 *digital* 數字部分而已。)

The semantic reanalysis of the term *Cyberport* has led Wu to believe that the existing norm *shu ma gang* 數碼港 is semantically inadequate and ambiguous because it covers only "half" of the meaning of the original term *Cyberport*.

In this case, a certain term conflict motivated by the evaluator's semantic interests is noted, and is followed by evaluation from which an inadequacy is established. The evaluation is made against the standard status of the consensus term *shu ma gang* 數碼港, with a reanalysis of the semantic content of the original concept. An adjustment is thereafter recommended for implementation

> "It would then be better to render *Cyberport* into *dian xin gang* 電信 (電訊) 港 (literally: telecommunication port), as this corresponds better to the semantic components of the term. (也許把 *Cyberport* 譯為電信 (電訊) 港更符合 *Cyberport* 的具體內涵。)"

The existing norm is evaluated according to a theory of semantics by analysing the degree of correspondence in content between the original and its translation, and also on the basis of the pragmatic function of the concrete act of translation. All these factors are immediately interest-based, arising from the evaluator's linguistic theories, and therefore normative in character and they determine both the translator's management strategy and the criteria used in the evaluation of the resulting translation.

Standardisation Bodies and the End Products

Professional Bodies

It is no coincidence that the development of Chinese IT terms in Hong Kong occurred, thanks to the interest of computer subject specialists. Subject matter and methodology develop when there is a need, and are pursued to the extent that they are the result of clear social needs. Like many other countries, technical terms, such as IT terms, are typically authorised by a professional organisation. In this regard, terminology concerns specialists of special purpose language.

Hong Kong Computer Society (HKCS)

Since 1995, the Hong Kong Computer Society has systematically been trying to further the use of standardised Chinese computer and IT terms to promote effective communications within the trade. Clear designation of translation equivalents at the institutional level, and the recognition of these names internally among specialists in the society, were excellent starting points. Agnes Mak, when elected president of the society in 1995, stressed that there was "a need to establish the standardisation of IT terminology database for Chinese" (*South China Morning Post*, 27 November 1995) and this need has since become one of the three essential goals for the society to achieve.

In 1994 the Hong Kong Computer Society won a contract from the Hong Kong Industry Department with the Department of Computer Science and Engineering of The Chinese University of Hong Kong (CUHK) to implement a database on a set of standardised Chinese computer and IT-related terminology. They focused on coordinating efforts made by several educational groups. The database was edited and revised by the China Computer Association (CCA) and endorsed by the China National Committee for Terms in Natural Sciences. The development of the project was funded by the Hong Kong Industrial Department, and sponsored by Sun Microsystems Inc., a technology solutions provider, and Sybase, Inc, an e-commerce applications provider.

The first phase of the Standard Chinese Computer Terminology Project was successfully carried out and the public could get access to the system through the HKCS website. With input from the China Computer Association and The Chinese University of Hong Kong, the Hong Kong Computer Society has developed an intelligent database for computer terms incorporating different names used in Hong Kong and China. The whole database serves to promote efficient internal communication within the trade.

Governmental Standardisation Bodies

In addition to being authorised by a professional body, technical terms are also typically authorised by a government body — an agency at the government level that tries to influence directly the territory's attention to the use of standardised terms. In Hong Kong the standardisation is being done separately by a number of government departments. These are agencies concerned with IT language problems and they produce standards for their own internal use, through the compilation of bilingual glossaries of IT terms. At this level, the broad target audience in the marketplace, that is civil servants and the public, is clearly identified. That it offers a broad direction can be regarded as terminology management at government level because it helps to coordinate all the internal language/terminology activities in relation to the use of Chinese IT terms, and define the general direction of language activities in which the departments are to be engaged — that is the translation activities in which an appropriate use of terms (i.e. a norm) is required.

Primarily there are three standardisation bodies within the government which carry out terminology work on Chinese IT terms, and they are:

(1) The Information Technology Services Department (ITSD)
(2) The Official Languages Agency (OLA)
(3) The Curriculum Development Institute Council, Education Department (CDIC)

(1) *The Information Technology Services Department (ITSD)*

As primarily functioned to "lead the Government and facilitate the community both in the development and the exemplary usage of information technology," according to its website, the goals of the ITSD include (1) promoting and enabling the extensive adoption and use of information technology in the Government; (2) enabling individuals, businesses and the Government to interact easily and securely through the use of information technology; and (3) promoting the wider use of information technology in the community.

In order to achieve these goals, the website reveals that the ITSD has produced products and services in order to be "compatible with IT strategy and meet evolving needs of departments and programmes." One of the services they provide to government bureaux is the compilation of a bilingual IT glossary. The glossary provides a reference of over 800 English-Chinese IT terms commonly used in the government. ITSD, when answering a question from the authors about the use of the glossary, claims that it is "a set of standardised Chinese Information Technology (IT) terms being used in the Government of the Hong Kong Special Administrative Region."

(2) *The Official Languages Agency (OLA)*

Another standardisation body is the OLA, one of whose goals, according to its website, is "to be recognized in the civil service as the authority on the use of Chinese and on translation between the two official languages." To achieve this goal they have developed various writing aids, reference materials and support services to assist civil servants either to promote the use of Chinese in their office, or to use Chinese in their work. They publish guidebooks on official Chinese writings, such as official correspondence, memoranda, file minutes, circulars, and notices. Their products also include twenty volumes of English-Chinese glossaries of terms commonly used in government departments. They cover different subject areas, including education, finance, meteorology, and trade and industry. In 1997 they produced the *Information Technology Glossary.*

The IT glossary contains some 2,000 entries on subjects related to information technology. The explanatory notes of the glossary it mention that the glossary "serves primarily to provide Chinese Language Officers with a handy translation aid and to standardise the Chinese translation of these terms."

(3) *Curriculum Development Institute Council (CDIC)*

Computer education in Hong Kong had a modest start in 1982 when Computer Studies was implemented in Secondary 4 and 5 as a measure to broaden the school curriculum. In 1987, Computer Literacy was offered to junior secondary students as well to provide them with some basic training in computer education. In 1992, AS-Level Computer Applications and A-Level Computer Studies were introduced along with other new sixth form subjects as a result of the new sixth form curriculum in Hong Kong schools. The syllabuses of the four computer subjects are designed to develop students' knowledge about computers and information technology skills including telecommunications.

To support teaching with Chinese as the medium of instruction and to facilitate communication and understanding, *An English-Chinese and Chinese-English Glossary of Terms Commonly Used in the Teaching of Computer Subjects in Secondary Schools*, was developed by the Computer Education Team within the Curriculum Development Institute Council and issued by the Education Department in 1999. The glossary provides Chinese translations of English terms commonly used in the teaching of Computer Subjects at all levels in secondary schools, and it covers IT terminology.

The Big Question: What Are the Actual Problems?

Indirectly, glossaries proposed by these professional or governmental authoritative bodies may exercise a strong harmonising influence on usage in special subject fields such as IT by the simple fact that a single glossary such as the one developed by the Information Technology Services Department or the Official Languages Agency may become the preferred terminological reference work for IT and its related subjects thereby establishing a virtual standard. Because of their leading position and role representing the government, such names can be seen as endorsement from authority and thus considered standard names regardless of the co-existence of any popular alternative names. With terminology work carried out by different bodies in different ways, not only the degree of consistency among these bodies becomes an issue that matters, equally important is the way they do the work, the principles they follow, as well as the methodology they adopt in carrying out the terminology work. The big question is whether these standardisation bodies know what actual discourse problems there are in the use of terminology, and what actual usage is.

From the glossaries prepared by the above four standardisation bodies, there was no mention about any criteria of terminological collection and the scope of its collection. It seems the selection criteria of what items to be included in an IT database differ from one to another. A terminological database should be a dynamic entity which is regularly undergoing changes as records are entered, completed, modified and deleted. How regularly are these databases kept up to date to reflect changes in usage? A clear distinction is being made between terminology as evidenced from usage in a diversity of pragmatic situations and the idealising tendency which sees a one-to-one correspondence between terminological and conceptual system. The top-down listing and implied prescription of terms can be understood in view of the development of the glossaries for very specific purposes by government agencies or departments in Hong Kong which thus have a directive rather than an advisory function. However, one of the modern attitudes to terminology work is that terms have linguistic variants in spoken and written language and within the same text type, so it is "essential to list these variants in descriptions of terminology." (Sager, 1990) This is nowhere mentioned in any of the four databases.

Government agencies have not issued any detailed descriptions of the procedures to be adopted in naming a new concept and making this name known. In Hong Kong there is little, if any, guidance about naming and even less about compiling glossaries. Government agencies regularly issue glossaries of terms as standards, but very few have firm guidelines for the selection, definition and publication of terminology. There is also great diversity between the glossaries. The perception that variation is a problem was also clearly demonstrated by the discussions of term problems related above. If standardised uniformity is the goal, the fact that variation is perceived as troublesome is of course a tautology. However, selection of the *one standard* term generates much intellectual concern with suitable criteria with which to evaluate the best candidate term.

What does not emerge and what in the authors' opinion should be the primary criterion is whether a term functions well in discourse. Once terms are in circulation, communication processes will reveal which terms are troublesome and require adjustment, even sanction removal when another is available. The process is that of discourse management through which language users note, evaluate, adjust, and as the case may be, implement adjustment of such language as is troublesome in subsequent turns of discourse. Those are the problems that should occupy the attention of term management agencies in their advisory capacity. Term management agencies are likely to be successful in their normative, standardisation capacity when they register unproblematic usage and bring representatives of specialist groups together to order and define that vocabulary in order to produce glossaries specific to that specialist group. For general IT "terminology," the language user is always right as long as communication flows without problems in understanding.

Notes

[1] Adapted from《詞庫建設通訊》(*Chinese Database Development Bulletin*), No. 17, September 1998, published by the Hong Kong Chinese Language Society 香港中文語文學會, and the《科技術語研究》(*Chinese Science and Technology Terms Journal*), Vol. 1, December 1998, published by the China National Committee for Terms in Sciences and Technologies (CNCTST) 全國科學技術名詞審訂委員會.

[2] CNCTST has provided valuable information and materials to this study, including a generous donation of two complimentary copies each of their quarterly journal, the《科技術語研究》(*Chinese Science and Technology Terms Journal*), published between December 1998 and June 2000.

[3] They include:《科技術語研究》(*Chinese Science and Technology Terms Journal*) published by the China National Committee for Terms in Sciences and Technologies 全國科學技術名詞審訂委員會;《中國科技翻譯》(*Chinese Science and Technology Translators Journal*), published by the Scientific Translators Association of the Chinese Academy of Sciences;《中國語文通訊》(*Chinese Language Review*), published by the Hong Kong Chinese Language Society 香港中國語文學會;《詞庫建設通訊》(*Chinese Database Development Bulletin*), published by the Hong Kong Chinese Language Society and《翻譯季刊》(*Translation Quarterly*), published by the Hong Kong Translation Society.

[4] The four participants include: Zhang Wei from the Computer Technology Research Centre of the Chinese Academy of Sciences, Tian Yujing from the Third Electronics Centre of the Information Industry Bureau, Yang Shiqiang of the Computer Department of Tsinghua University, and Lin Jian from the

Computer Centre of the Chinese Academy of Sciences. In《科技術語研究》
(*Chinese Science and Technology Terms Journal*), No. 1, March 2000.

References

Cabré, M. Teresa (1999). *Terminology: Theory, Methods, and Applications.*
Translated by Janet Ann DeCesaris. Amsterdam and Philadelphia: John
Benjamins Publishing Company.

Fang, X. D. (1999). "Between Names and Spirits."《科技術語研究》(*Chinese
Science and Technology Terms Journal*), No. 5, pp. 10–11.

Jernudd, B. H. and J. V. Neustupný (1987). "Language Planning: For Whom?" in
Lorne Laforge, ed., *Proceedings of the International Symposium on Language
Planning*. Ottawa: Les Presses de l'Université Laval, pp. 69–84.

Jernudd, Björn H. and Jiri V. Neustupný (1991). "Multi-disciplined Language
Planning," in David F. Marshall, ed., *Language Planning*. Amsterdam and
Philadelphia: John Benjamins Publishing Company, pp. 29–36.

Jernudd, Björn H. (1997). "Theoretical and Practical Dimensions of Language
Planning Work," in *Proceedings of the European Conference on Language
Planning, Barcelona 9–10 November 1995*. Barcelona: Generalitat de
Catalunya, pp. 9–19.

Sager, Juan C. (1990). *A Practical Course in Terminology Processing*. Amsterdam
and Philadelphia: John Benjamins Publishing Company.

Wang, H. D. (1999). "Reviewing Translation for the Sake of Standardising
Information Media Terms — An Introduction to the Translation Standardisa-
tion Committee for the Chinese Media, Singapore."《科技術語研究》(*Chinese
Science and Technology Terms Journal*), No. 4, September.

Wang, Ning (2000). "Globalisation, Cultural Studies and Translation Studies."
Translation Quarterly, No. 15, March, pp. 37–50.

Wright, Sue Ellen and Gerhard Budin (1997). *Handbook of Terminology
Management*. Amsterdam and Philadelphia: John Benjamins Publishing
Company.

Wu, Wenzhao (1999). "The Cyperport and the Language Gap." *Chinese Language
Review*, No. 61, October, pp. 33–34.

Terminological Problems and Language Management for Internet Language Professionals in Hong Kong

Charlotte To and Björn Jernudd
Department of English Language and Literature
Hong Kong Baptist University

Introduction

The incredible development of the Internet presents today's companies with the opportunity to reach around the world to find new markets for their products. In order to grasp this chance, many companies develop their own websites to make access for their customers easier. Since the development of the web industry has been so rapid, the number of new words for new concepts in technical computing and acronyms have been increasing immensely. Due to the growing internationalisation of business, new English terms are exported to other countries where English is not the usual language. The technical computing terminology causes communication difficulties among Internet language professionals, in particular for those translators, web editors and technical writers who do not have much knowledge of technical terminology.

These difficulties also exist in Hong Kong. Therefore, the following questions and concerns are timely:

(1) What are the terminological inadequacies and problems actually encountered by Internet language professionals in Internet business firms in Hong Kong?

(2) In what situations do they encounter these terminological problems?

(3) What adjustments are made for those term inadequacies that occur online in the process of discourse?

(4) What solutions are found for those term problems that are referred offline?

This paper will collect and report on Internet language professionals' first-hand experiences with term problems in an Internet business in Hong Kong.

Model of Individuals' Management of Language in Discourse

A theory of language problems (Neustupný, 1973) explains the relationship between discourse and people's behaviour towards discourse. The theory of problems focuses on how participants in the process of language management note and evaluate features of language and of language systems. (Jernudd and Neustupný, 1987) The language management model postulates that in discourse participants:

(1) *produce* utterances (in written language, sentences);

(2) *monitor* the language that constitutes these messages, *note* a potential deviation from norm (which includes a gap, say, of vocabulary);

(3) *evaluate* the deviation from norm in the noted item, thus determining whether the noted deviation constitutes an inadequacy;

(4) *select* an adjustment;

(5) *implement* the adjustment by (re)producing the utterance (rewriting the sentence).

The discourse management process is interactive between self and other; and it may stop at any step in the modelled process. It is ever present in any communicative act and enables continued communication. The demand for precise and untroubled specialist communication makes it rather more likely than in general conversation that overt attention will have to be paid to terms. New vocabulary requires dissemination and collective, possibly authoritative, endorsement as terms, and current in-group usages require continuous affirmation of the constancy of their forms and definitions. Thus, *term problems* arise when individuals bring up noted deviations for deliberate attention in the discourse community, thus *talking about* their specialist language. They engage in meta-communication. Specialists may, for example, be unable to agree on an adjustment to an inadequacy that both recognized in their discourse. The adjustment may concern how to express a new concept, how to disentangle an abbreviation, or how to render an English term into Cantonese. They therefore ask colleagues for help, look up sources, or, depending on the felt severity of the issue and on the degree of organisation of language management of terms for the particular discourse community, they may refer the matter to the relevant term management organisation.

Data Collection

Two opposite forces govern term management in today's Internet industry in Hong Kong. On the one hand, Internet language professionals' communications require explicit and stable agreement on the expressive forms of terms and on their definitions to capture and communicate concepts precisely. This process is not supported by organised language management in Hong Kong but is for the moment largely confined to collegial consultations in the workplace. On the other hand, they seek new knowledge and develop new practices, therefore running into discourse problems with existing terms and definitions. Therefore the need arises for the coining of new ones.

Two types of data have been collected in this study: (1) records of potential deviations by asking volunteer participants to keep a record of queries that arose during their work; and (2) transcription of a focus group discussion, conducted for confirming the evidence that was collected from the records. The subjects, three males and two females, aged from twenty-five to twenty-eight, were volunteer participants. They are the web editors from two local Internet Content Providers (ICP). These two ICPs were launching two portals, both with English and Chinese versions. Web editors were chosen because they are mainly responsible for writing and editing the contents of these websites. However, they are not very familiar with technical computing terminologies, compared to other employees of the ICP, such as web designers and web developers.

The Records

The subjects were requested to record any doubts, difficulties and queries concerning technical computing vocabulary as they came up at their workplace, in written or spoken, English or Chinese, discourse. Records were kept for about one month. Each subject was given 31 record sheets based on the sample query record of the Catalan Language Service, University of Barcelona. (Cabré, 1999:152) The record sheet is a one page piece of paper with spaces to be filled in for the subject's name and title, the date; to write in the noted item and details on the situation in which the noting occurred.

Focus Group Discussion

A half-an-hour focus group discussion was conducted right after the subjects had finished record-keeping. It was divided into two parts, the individuals' term and discourse management and corporate discourse management. The first author guided the discussion. The discussion questions are reproduced in the Appendix. The discussion was unobtrusively taped and transcribed.

Findings

Records of Query

There was a total of 80 records received from the subjects. In the following, each vocabulary item that was noted so as to produce a record is called a query. We will account for the queries by grammatical categories.

Grammatical Categories of the Queries

Most queries were nouns (92.5%). 30 out of 80 collected queries were acronyms. Only 6.25% of the queries were verbs and one was an adverbial. Not surprisingly, most queries concerned the English language (97.5%), only 2.5% concerned Chinese.

Most queries would be regarded as terminological problems. The records show that the subjects consulted others or consulted terminological sources in order to solve the problems that had arisen. This may be a result of the highly conscious manner in which the data was collected but does not in our opinion invalidate the items. Observation or prior training of record-keepers would have produced many more queries and among them many more discourse inadequacies that would have been adjusted by the subjects themselves.

In the immediately following sections, we offer selected examples of terminological problems.

Example 1

The noted item: *[siŋ]*
Situation:

In a project meeting, the marketing director and the web editor asked the programmer for details about the database of a project. The programmer replied that two database systems can be [siŋ] by making use of some programming methods. This was the first time the web editor heard this word and she queried it with the marketing director after the meeting.

Adjustment:

When the web editor consulted a colleague on the noted item, she learnt that [siŋ] is short for "synchronize." While a linguist can speculate, after the fact and with knowledge of the source, how this short form was created, the form itself does not reveal this. A linguist would say that the original word "synchronize" has apparently been shortened to [siŋk] and then [k] dropped (and we may understand this latter process either as assimilation to the preceding nasal or a dropping out of the final C [k] of two C's [ŋk] to conform to Cantonese syllable structure, in both processes under the influence of a speaker's Cantonese). Once [k] has been dropped, it cannot be retrieved "by rule." [siŋ] is not transparent as a short form at least. Therefore it was noted and the web editor initiated adjustment by another person.

Example 2

Noted item: *netvigation*
Situation:

A web editor noted that "netvigation" occurred many times in drafts of reports, agendas and minutes to refer to different sections and sub-sections of a website, evaluating the noted item negatively.

Adjustment:

The web editor reported that she regarded "netvigation" as a deviation from "navigation" and called it a misspelling. The adjustment would therefore be "navigation." However, she did not overtly correct the documents nor discuss the item with others.

Example 3

Noted item: *ASP*

There are two entries for "ASP" from the same record-keeper, a web editor. In the first entry, the web editor reported that she heard "ASP" in a sales meeting and was not sure what it referred to. In the second entry, the web editor reported that a programmer had recommended a solution for "ASP," a new project while answering e-mail for technical advice. The web editor was puzzled to note that this "ASP" must have a different meaning.

Adjustment:

She theorized that the noted "ASP" could be an abbreviation and therefore asked a colleague after the meeting what it abbreviates. She learnt that "ASP" stands for "Application Service Provider" (which would be cumbersome to insert in an otherwise Cantonese utterance in which "ASP" was used). The web editor checked the other expression with the programmer and learnt that it stands for "Active Server Pages," thus confirming her guess that this "ASP" is different.

Example 4

Noted items: *plug-in, web-based tool*

A web editor and her colleagues disagreed on using the terms "plug-in" and "web-based tool." They checked with a programmer in the same company.

Generally speaking, the meanings are similar. However, "plug-in" is a kind of "web-based tool." Adjustment establishes that the latter is the super-ordinate concept and the term can therefore be used in place of "plug-in," but the latter is the more specific term and should be used when specificity is desired.

Selected Problems Reported from the Focus Group Discussion

We reported on selected problems by reproducing transcribed utterances in the focus group that discuss language problems and their solutions, annotating the utterances as to their function in language management: whether the utterances reveal noting (and of what trouble-item), evaluation, or adjustment, and whether this is done by self or by other.

Example 5

An *acronym*: SKU
Transcript:

> The most impressive one should be the SKU one … When we, *The Avenue*, were having a summer sale promotion recently, I always urged my colleague

to confirm the details with me. She was so busy that she didn't make up her mind until very late. She later on told me that there were around 50 SKU designated for the promotional sale. So I told my man about this. [Noting and evaluation:] *He didn't understand what SKU meant and* [Initiation of adjustment:] *asked me what that was.* [Adjustment:] *I then told him that the full name of SKU was Stock Keeping Unit, and that we'd focus on one category which included something like napkin, handbag. I said there were around 50 units of products as they said 50 SKU.*

Example 5 continued: different pronunciations of SKU
Transcript:

However, [The problem:] *not everyone pronounces the word SKU in the same way.* Some people might pronounce it like "skill" and I would not be aware of the word. I think that acronyms should not be pronounced as a word. Instead, that [Suggested solution:] *they should be pronounced separately by letters like, S, K...*

Example 6
Definition of a *term*: site map
Transcript:

Taking the site map case as an example. From the viewpoint of a designer or a programmer, a site map should be made in that way. But to us, a site map can be as simple as a handwritten rough draft. [Noting and evaluation:] *We had very different perceptions towards the term site map.*

The subject points to the problem that different people on the same project in the same firm assign different meanings to a specialized vocabulary item. This is probably a result of previous misunderstandings.

Example 7
A *hyponym*: registered user, unique user
Transcript:

S1: [Bringing up a language problem:] I've been corrected by a colleague once recently. I was discussing if there is a unit in use for measuring the number of registered user so as to reflect the value of the portal website. [Other-initiated adjustment:] *A colleague told me to use the term "unique user"* [Other-initiated noting and negative evaluation:] *instead of using "registered user."*

S2: It reflects no value if you use "registered user" to do the measurement. Many visitors may use various identities to register as users of the web site. One visitor can register for as many times as he wishes!

Both expressions are subordinated to "Internet user" but their definitions differ.

Situations in Which the Queries and Problems Arose

One of the research questions was to find out in what situations the subjects encounter terminological problems. Queries arose in the following situations at their workplaces (the number in the bracket indicates the number of people reporting that situation):

(1) casual conversations/phone conversations with colleagues in the office (32)
(2) attending business meetings (31)
(3) reading/preparing business documents and proposals (5)
(4) reading Internet magazines/journals/articles (5)
(5) Internet search (for work purpose) (4)
(6) reading/replying to business e-mails (2)
(7) using software/a special computing programme (1)

Subjects mostly encountered the terminological problems during business meetings and (casual/phone) conversations with colleagues in the office. In order to prevent communication breakdown, the subjects adopted various adjustment strategies, which are shown in the following section.

Adjustment Strategies for the Terminological Problems

Subjects reported three kinds of noting and evaluating inadequacies:

(1) self-noting and evaluation of inadequacy
(2) inadequacy noted and evaluated by others
(2) noting and evaluation of others' inadequacy

Self-noting and evaluation usually occurred when reading a manual or magazines, using software, Internet search, attending a seminar, during talks and briefings, or preparing documentation. Others were involved (quite naturally) in situations of two-way communication such as in cooperative work, business meetings and casual conversations in the office.

The following adjustment strategies were adopted for each of the above cases (the number in the bracket following each strategy indicates the number of people selecting that strategy):

(1) Self-noting and adjustments of inadequacies:
    a. by asking/discussing with others
       – immediately (34)
       – after a certain period of time (6)
    b. by checking a bilingual dictionary (1)
    c. by asking/discussing with others and looking up a dictionary/ other reference (5)
    d. by guessing the meaning of terms from context (2)
    e. by guessing the meaning of terms from context and asking others (2)
    f. by checking other related subjects reference/materials/Internet (6)

(2) Inadequacies were noted and adjusted by others:
    a. noting without any adjustment strategies (2)
    b. noting with an explanation given (8)
    c. noting with the correct term given (no explanation) (1)

(3) Noting and adjustment of others' inadequacies:
    a. noting without any explanation given (1)
    b. noting with an explanation given (7)
    c. noting with the correct terms given (no explanation) (3)

A majority of the subjects noted the terminological problems by themselves. They preferred to ask or discuss the problems with others immediately or after a certain period of time. In fact, this is the quickest and most convenient way to solve terminological problems.

Management of the Choice of Chinese and English Terms in Chinese Discourse

All the subjects responded that since they have learned and used English terminology, in most cases, English terminology is their first priority. In spoken Chinese discourse, code-mixing is very common. As in written Chinese discourse, common practice is:

    a. simply use the English term, with Chinese explanation
    b. use the Chinese term, but glossed with the English one

> *c.* simply use the English term without any Chinese term or explanation

Interestingly, using pure Chinese terminology in any discourse does not seem to be welcomed by the subjects. From their viewpoint, Chinese terminology causes comprehension difficulties. They have to recall the English equivalents in order to understand. An additional concern is differences between Chinese terminologies for China and for Taiwan. (Chiu and Jernudd, 2000)

Use of Language Resources

The subjects use language resources in order to make adjustments. The most common language resources that they use are:

> *a.* bilingual dictionaries
> *c.* online glossary lists
> *d.* reference books
> *e.* Internet magazines and journals
> *f.* Intranet
> *g.* user menu in the form of a CD-ROM
> *h.* seminar/talk/briefing organized by the companies

Term Management in the Firms

According to the subjects, few language resources were provided by the companies (although the list below is long). Subjects prefer using their own resources. The following are examples of language resources provided by their companies:

> *a.* user menu in the form of CD-ROM or paper
> *b.* documentation
> *c.* memo/e-mail
> *d.* video
> *e.* Intranet
> *f.* legal consultant (only for drafting contracts or other legal issues)
> *g.* seminar/talk/briefing for the staff
> *h.* training course (not so common)

There exists no particular language policy or language department in the companies which aims to support the employees on language issues.

However, subjects would welcome such facilities. Some of the subjects even suggested an enquiry hotline for answering term problems or other language problems that arise at work. This would save a lot of searching time for them to find out appropriate adjustments.

Discussion

Terminological Problems Encountered by Internet Language Professionals in Internet Business Firms in Hong Kong

Categories of terminological problems included acronyms, pronunciations, spellings, meanings and usage. In particular, the use of acronyms causes difficulties because the concepts behind these acronyms were still new or unknown. Problems of homonymy, synonymy and hyponymy also arose. In order to master related problems with technical terminology, Internet language professionals have to first understand the concepts behind the terms.

Internet language professionals in Hong Kong also encounter another macro-problem: Chinese technical computing terminology. Hong Kong is professionally a bilingual city. Using English terminology in Chinese discourse (both spoken and written) is common. Internet language professionals learn the English terminology first. Since they do not have a need to learn their Chinese equivalents, inadequacies are likely to arise when they first come across Chinese terminology.

Situations in Which Internet Language Professionals Encountered Terminological Problems

Most terminological problems were encountered in "casual conversations with colleagues" and "business meetings." Perhaps in these situations people with different backgrounds meet. In other words, the participants bring different technical terminology into the discourse. Therefore, more deviations become possible compared to interaction in other situations such as in close cooperation in a project group.

In fact, the situations do impose some constraints on the adjustment and implementation process. Situations such as business meetings may prevent adjustments taking place. Owing to embarrassment, the presence of VIPs of the company, the priority to attend to the progress of the meeting, and so on, participants hesitate to request or offer adjustment.

These external factors should be taken into account in evaluating individuals' term management. People remain ignorant sometimes because of the force of such external factors rather than because of lack of adjustment strategies. It follows that meetings should allow room for questions on terms just as on other matters of substance.

Adjustments and Solutions

Many adjustments in the full data set were simple corrections during discourse, i.e., giving a correct term or by offering an explanation in the flow of an ongoing discourse. The findings match what Jernudd and Neustupný (1987:76) have written:

> A complete process of language management starts from marking in discourse of some aspects of that discourse as inadequate, finished with the implementation of the simple correction design, again in discourse. Simple correction in discourse is thus the most important category of the entire language management process.

Unlike adjustments during discourse, but closely allied with it, the most popular solution to an overt language problem offline was asking and discussing the problem with others immediately. This kind of adjustment could also be regarded as a side-sequence with the inadequacy as its topic during ongoing discourse.

In the focus group discussion, similar comments were found. In addition to seeking assistance from experts, colleagues or friends, some subjects even suggested having an enquiry hotline for answering questions on terminological problems. The data collected in the focus group discussion support findings from the data collected in the record sheets.

Asking and discussing immediately as a query arises with someone who may know the solution (an adjustment) is very probably the most cost-effective way to adjustments in communication.

Similarly, among offline adjustment strategies, seeking assistance or advice from someone knowledgeable seems to be the most popular one.

Significant Effects of Term Problems on Business

Inadequacies have negative effects. For example, our data shows that when terminological problems occurred during business meetings, people would spend meeting time to explain and discuss the meaning of terms, or would require time afterwards, even a renewed meeting in one case, to sort out the language problem. In other words, a meeting can be derailed and this costs money. Even more costly would be the situation when after long discussion, no agreed norm or solution has been found.

Conclusion

Internet language professionals encounter terminological problems. Terminological problems include technical computing jargon, especially the prolific acronyms. Inadequacies arise in both spoken and written discourse in terms of pronunciation, spelling, meaning and usage.

Subjects encounter terminological problems in a number of situations, including but not limited to, business meetings, casual or phone conversations with colleagues in the office, seminars, talks and briefing sessions, reading technical manuals, preparing business or legal documents, drafting articles for publication on the web, and using software.

The subjects reported that Chinese terminology causes difficulties to comprehension. The subjects have to recall the English equivalents to understand them with the help of context. In order to avoid difficulties with terms in Chinese, they generally gloss in the other language or use an English term in an otherwise Chinese discourse.

The subjects use a variety of adjustment strategies. Offline, seeking assistance or advice from someone seems to be a common strategy. The results of this study suggest that interactively seeking adjustments immediately an inadequacy arises with someone who may know is very probably the most cost-effective, under present circumstances. It is significant that subjects supported the notion of a hotline to help with solutions to language problems.

Another possibility is to bring together Internet firms, terminologists, and subject experts to agree on a norm for technical Internet terminology, i.e., to *standardise* Internet terminology in Hong Kong for the sake of firms in the private sector. This raises the question which should be the authoritative standardisation body and how to get Internet firms actively involved.

This brief study may reflect practices in only a fraction of the entire Internet industry, yet the authors are confident that the issues it raises are general ones. In any case, further research must be conducted so as to reveal more fruitful and reliable facts about language use and language problems in the Internet industry in Hong Kong.

References

Cabré, M. Teresa (1999). *Terminology: Theory, Methods and Applications.* Amsterdam and Philadelphia: John Benjamins Publishing Company.

Chiu, Aman and Björn H. Jernudd (2000). "Chinese IT terminology Management in Hong Kong." Paper presented at a conference on "Translation and Information Technology" organised by the Department of Translation at The Chinese University of Hong Kong, on 8 December.

Jernudd, Björn H. and Jiří V. Neustupný (1987). "Language Planning: For Whom?" in Lorne Laforge, ed., *Proceedings of the International Symposium on Language Planning.* Ottawa: Les Presses de l'Université Laval, pp. 69–84.

Neustupný, Jiří V. (1973). "An Outline of a Theory of Language Problems" (First draft). Paper prepared for the section Language Planning, VIIIth World Congress of Sociology, Toronto, 16–24 August 1974.

Appendix

Questions for Focus Group Giscussion.

Part 1. Individual Discourse Management (Terminology)

1. What kind of language problems do you usually encounter at your workplace, in particular terminology?
2. Usually under what situations? Any examples?
3. One way communication (e.g. reading) or two ways (e.g. dialogue)?
4. Chinese terms or English terms?
5. What is the result then? Miscommunications or communication breakdown?
6. Who note these language problems? Yourself or others?
7. Would you evaluate it (i.e. how incorrect is it) and then correct it right away after you have found the problems? Why or why not?
8. If yes, how and where did you find the adjustment resources? What are these resources? If no, why not?
9. How about if you note that others have the language (or term) problems, would you correct them? How would you help them?
10. If others note that you have the language (or term) problem, would they correct you? How would they help you?
11. What kind of adjustment strategies do they (or you) usually give you (them)?
12. How would you (they) use (implement) it?

Part 2. Corporate Discourse Management

1. Any in-house management of terms or in-house text production (e.g. glossary) in your company or department?
2. If yes, what are they? How is that organised and distributed?
3. If not, do you think there is a need to have that? Or do you want to have that? Why?
4. Are there any language policies in your company or department? What are they? Which body within your company or department authorised it?
5. If not, do you think there should be some language policies to help and support the employees?
6. Any language training/resources does the company offer and support you? If yes, what are they?
7. Do you think that training is enough or appropriate? Why?
8. If not, what kind of training do you think the company should offer? Why?

PART 3

Critique and Training

The Internet and Other IT Resources: Tools for Translators within a Translation Programme

Beverly Adab
School of Languages and European Studies
Aston University

Introduction

In a recent paper in *Target*, Li Defeng of The Chinese University of Hong Kong published results of a survey of professional translators and their perceptions of how well their training had prepared them for the real world of translation. In this paper he makes the following observations:

Courses that students found most helpful included:

a.	Translation projects	19.%
b.	Translation theory	14.3%
c.	Specialised translation courses	30%

Areas where students felt best prepared:

a.	English and Chinese competence	57%
b.	Translation skills	38%

Areas where students felt least prepared

a.	Subject matter knowledge	57%

Professor Li also reported the following as being key problems experienced by new translators:

   a. Subject matter knowledge;
   b. How to select appropriate register;
   c. How to apply the correct textual format;
   d. How to select the correct Target Language (TL) terms;
   e. How to choose the most appropriate TL style for the text type;
   f.   Where to find relevant references;
   g. How to write in correct style and standard Chinese;
   h. Making more effective use of the Internet;
   i.   Developing common sense strategies for dealing with these problems.

Other problems:

   a. Time constraints;
   b. Language competence.

Meeting Translators' Training Needs

These findings could quite clearly be read as indicating that trainee translators feel a need for more practice in Translation for Special Purposes (TSP) and they are aware of the need for specialist knowledge in a range of domains of professional activity.

We would argue that the most appropriate way to meet these needs may not in fact be to offer more and more courses in specialist subjects. This may be possible in situations where a translator is only likely to work in one or two specialist areas. What about, however, the more typical translator profile, someone who needs to earn a living, who needs to be flexible, adaptable and to be able to tackle a range of texts on different subjects? How can the university translation programme prepare trainees for this and for other demands of the professional environment. These are the questions that we will address in this paper.

Structure of the Paper

We will begin by reviewing briefly some of the key demands of the professional translation environment. We will then consider what determines whether or not a translation is useful for its intended purpose and what is the nature of the type(s) of translation competence required to achieve this. We will touch on the role of different contributory aspects of the translation programme, including translation theory but also an introduction to IT and to the use of electronic corpora (comparable and parallel) within the programme. We will make reference to the under-graduate and postgraduate programmes in translation studies at Aston University, to demonstrate how we aim to combine theory with practice in order to equip students with the relevant analytical and documentation skills, so that they are competent to work in a variety of subject areas on a range of text types and still produce a functionally adequate translation.

The Real World Translation Environment

As we are all aware, in the real world of those who are gainfully employed, IT skills are no longer optional but a prerequisite for many professional posts. Information is needed, at speed, from the most up-to-date sources, if those in the business environment are to succeed. Transactions and communications take place in electronic form, almost, if not actually, in real time. We have to try to keep up, or else get left behind. In the professional translation market, the demands are many but the driving force is simple. Translators have to produce translations which are adequate for their intended purpose. As Nord states (2000):

> A translation can be considered *functional* when it fulfils the intended communicative purpose as defined by the client or commissioner in the translation brief.

This view is supported by Vermeer and *skopos* theory, which prioritises the translation brief and the target reader's needs in relation to this brief, as the determining criteria in the decision-making process. Vermeer (1996) argues that to prioritise the *skopos* is to expand the possibilities of translation, to increase the range of possible translation strategies and to free the translator from the restrictions of a (mainly meaningless) literal approach. It also emancipates and empowers the translator, who becomes accountable to the target reader as well as to the Source Text (ST) author and to the translation commissioner.

To ensure that, as professionals, they produce texts which fulfil their intended purpose, translators have to possess a wide range of skills, for which they need adequate training and preparation. Translator training institutions and academic programmes cannot possibly equip translators with every imaginable variation of these skills, they can only provide a solid foundation in different aspects of translation competence, or sub-competences. Chesterman (2000) affirms that, "Adequate translations are the result of just the right configuration of the five competences." For Chesterman and other contributors to this volume, these five/six competences relate to:

Language competence — near-native competence in the TL and sophisticated command of mother-tongue;

Textual competence — translators have to internalise culture-specific norms of text production and to be sensitive to both differences and similarities between sets of norms across languages and cultures;

Subject competence — translators can be taught to conduct research and to identify key concepts for any domain; also to know when the domain is too complex or the content of the text too specialised for a translator to produce a functionally adequate text;

Cultural competence — culture-specificity of ideas is a property of pragmatic as well as literary texts. An essential component is a comparative and contrastive awareness of similarities and differences between source and target cultures, with reference to patterns of thought, modes of behaviour, value systems and taboos, communication processes and reader expectations. Knowing what Nord (1991, 1997) describes as the "rich points" which highlight the differences between the two cultures is key to the production of an adequate target text;

Transfer competence — translation is all about transfer of information, in one form or another. Learning to identify translation strategies and how best to use them is part of the overall transfer competence;

Translation competence — a combination of all of the preceding competences.

Anderman and Rogers (2000) note that:

> The emergence of new technology has drastically changed the working environment of professional translators. As a result, translation competence can no longer be defined in isolation but must be viewed in relation to the requirements of a rapidly developing information society.

In the limited time available to academic institutions and within the context of the need to offer not just a vocational training, but also and equally importantly, a serious programme of academic study, certain questions must be anwered:

   (1) How much can the academic translation programme hope to achieve in terms of preparing students for the IT demands of the

profession, whilst not neglecting the more formal, theoretical aspects of the programme?

(2) How can we combine translation theory as a discipline, with experience in the process of translation and competence in translation as product?

At Aston University, we are trying to incorporate the use of information technology into our undergraduate and postgraduate translation programmes.

Translation Theory: Principles Underlying Our Programmes

We adopt a functionalist, pragmatic perspective to our applied translation courses, to encourage the view that theory and practice are interdependent, in that practice relies on knowledge of theoretical concepts to improve or enhance performance, whilst theoretical perspectives are based in part on observation of performance in practice. We encourage students always to demand a translation brief, to adopt a professional methodology for text comprehension and text production, and to relate their decision-making to aspects of translation theory. In addition, we expect students to conduct a thorough Translation Oriented Source Text Analysis (TOSTA) as advocated by Nord (1991, 1997), for each source text.

Aston Translation Programme — Undergraduate Studies

All students follow the core courses in language competence and cultural competence which form part of the standard modern languages degree programme at Aston. In addition, translation students follow specially-designed courses in translation competence.

In the First Year students follow a course entitled, "Introduction to Basic Concepts of Translation," which includes lectures on key aspects of translation theory and the obligation to apply these concepts in translation performance, for source text analysis and target text production. This is supported by courses in linguistics and comparative grammar.

Our emphasis on comparative and contrastive knowledge is further reinforced through training in the importance of text-type conventions and norms of behaviour. In the Second Year, students follow a course entitled "Intercultural Text Comparison," which focuses on similarities and differences in text typological conventions for a range of text types.

Comparative and contrastive knowledge for text-type conventions also includes awareness of the role of terminology in translation. This is another Second Year course. We do not aim to train terminologists, but to ensure that our students are aware of how to construct, use and evaluate a terminology for a domain-specific translation task. This involves an introduction to the role of corpora in translation and in the production of a terminological database. Courses in semantics and sociolinguistics are also offered.

The Third Year is spent in the country or countries of the target language(s), in a university translation programme or in a translation-oriented work placement.

In the Final Year, students follow a further course in contemporary translation studies, complemented by, "Professional Translation, Theory and Practice," and a research module on "Translation for Specific Purposes;" the former is assessed by essays, the latter by the production of a translated text supported by evidence of the application of theoretical concepts.

Use of IT in the Programme

All assessed work has to be submitted in word-processed form. Translation students are expected to use and to cite a wide range of sources, including electronic sources and the Internet, for corpus compilation and to support translation decisions.

First year assignments will include an introduction to all available tools for the translation, including using the Internet, electronic reference sources and CD-ROMS. Students are introduced to *Multiconcord* and to the basic principles of how machine translation and translation memory programmes (such as *TRADOS*) operate. Students are required to use Machine Translation (MT) programmes such as *Babelfish*, available on Alta Vista, to evaluate how functionally adequate the target text may be and why, again with reference to theoretical concepts of translation. Using such readily available MT software is also a good introduction to proofreading and post-editing, as well as generating a framework of guidelines for text production and self-monitoring.

For Terminology in the Second Year, students are shown how to construct a simple table (based on pre-determined fields) using a standard *Word for Windows* programme, also how to use the find/search tools to check and use their own terminology.

In Final year, students have to gather texts to form a corpus for reference purposes — very often the primary source for these is Internet searches, with selection according to pre-determined criteria for inclusion. They also search for papers by translation studies scholars available via electronic as well as printed sources — this practice is becoming ever more widespread and provided the source is reliable and authoritative, the Internet can complement quite well other existing printed sources of scholarly writing.

Aston Translation Programme — Postgraduate Studies

In the Masters Programme we continue our emphasis on use of corpora and the interdependence of theory and practice. For this programme we focus on translation in the European Union, as this environment provides a case study for the particular features involved in a multilingual text production environment. It also allows students to develop awareness of particular types of constraints on the translation process and to test their understanding of how a functionalist approach can be applied in different situations. Study of the European Union, its development, institutions and primary practices, gives contextual knowledge and contemporary translation theory is revisited.

The Translation Environment in the European Union — Constraints and Demands

(1) A multilingual environment — texts are produced by committees of native and non-native speakers, in one or two working languages, for translation into the other working languages.

(2) There is a need for precision, clarity, lack of ambiguity — texts may be legal documents and have to lend themselves to a single possible interpretation, so that translated versions can be equally clear and unambiguous.

(3) The European Union has clear guidelines on text production, relating to all aspects of this: use of titles, inter-textual references to other documents, decisions; terminology for institutions, official titles and roles, directives, key concepts, even collocative use. These guidelines also regulate layout, requiring identical text sections to appear on each page, for ease of comparability between different language versions of the same document.

All of these constraints have led to the development of a very particular use of the key working languages, particularly of English. This "Eurospeak" tends to use a reduced repertoire of language items and to generate collocations and combinations unknown to native speakers in the same language as used within its source culture. It has become a hybrid form of the language, containing examples of adaptation of forms from other languages, some of which may well become accepted new forms, others of which continue to be seen as interference and contamination. The tendency is to move towards the lowest common denominator, to demand simplicity of structure in order to ensure clarity for non-native speakers and to sacrifice niceties of style on the altar of the *lingua franca*. There is no need for explicitation or adaptation, insertion or omission, since European Union culture-specific knowledge will be common to all readers.

All of these constraints have also led to a restriction of the freedom of the translator, yet paradoxically, they may also be a surer guarantee of functional adequacy, since the strategies to be adopted are pre-determined to a large extent and the scope for selection, hence for possible inappropriate choices, is necessarily less open. So the translator's expertise in this domain may be described in terms of a comprehensive knowledge of all aspects of this multilingual culture of text reception and a respect of the text type norms for this context — of course, in addition to the relevant language skills.

However, simply to train a translator for this environment would be unfair and restrictive, like the example of training someone to drive *only* a particular make or model of car. What we try to inculcate is an awareness of how to identify the important features and constraints of any translation situation, also how to adapt translation behaviour accordingly and use all relevant sources.

How Trainee Translators Can Be Prepared for the Rapidly Evolving Translation Market

During the last couple of decades, the emergence of new technology has drastically changed the working environment of professional translators. As a result, translation competence can no longer be defined in isolation but must be viewed in relation to the requirements of a rapidly developing information society. How then *are* academic institutions facing up to the challenges posed and what steps are they taking to prepare future translators for this brave new world?

This need for skills of adaptation to the demands of the environment is achieved in part through practice in translation of a range of text types from different domains, including technological advances (computing, mobile telephones etc); business documents (company reports, marketing and advertising texts); instructions for use (everyday products); and printed press articles (editorials, letters, interviews, comments on social issues).

As has hopefully been demonstrated, we encourage students to see the systematic collection of authentic examples of comparable and parallel texts as an integral part of the pre-translation process. They are required to support their translation decisions by reference to the texts collected and the end of course assessments also reflect this requirement. We expect students to develop critical awareness and to be able to evaluate for themselves any target text, supported by reference to authentic, parallel texts.

Corpora in the Translation Programme

Principles and Practice

The key principles which our students are expected to acquire include:

(1) The types of corpora that can be useful for translation purposes: parallel texts (ST/TT) and monolingual comparable sets of texts on a given subject/of a given type;

(2) The uses of corpora: through an inductive process of information gathering and use of texts collected for practice and for assessment: identification of text typological features by means of descriptive comparative and contrastive analysis: awareness of similar/different features; the compilation of domain-specific knowledge and terms — as well as other uses for more linguistically-focused study;

(3) The importance of reliability and representativeness of a corpus — in relation to pre-determined aims and criteria for selection;

(4) Speed of search and ease of access in the translation process — meeting deadlines;

(5) Intended use — for quantitative analysis or for more detailed, qualitative study;

(6) Synchronicity or diachronicity — the different uses for such corpora;

(7) Developing a systematic methodology for text analysis;
(8) Evaluating the reliability of an existing TT and its appropriate-
     ness for the stated purpose — extending this critical ability to
     evaluate one's own work;
(9) The computer as a tool and resource;
(10) The use of language in a multilingual environment — European
     Union case studies.

Using Corpora in the Academic Translation Programme

Bowker (1998) discusses how using corpora can be of use in a translation
programme, in order to meet the different demands placed on (trainee)
translators. She notes that these demands include subject knowledge for the
domain, source language competence and target language competence.

We agree with Bowker that if professional translators cannot, and
quite rightly, be expected to have domain-specific knowledge in a wide
range of domains, even less so can this be expected of students in an
academic translation environment. The primary communicative function
of domain-specific texts is usually the informative function. What is
needed is competence in documentary research in order to acquire
sufficient understanding of subject of the text and thus to allow an accurate
transfer of information. This will involve comparable monolingual
corpora, to allow understanding and knowledge of key concepts and
appropriate terminology in both SL and TL. Compiling such corpora also
requires students to develop and then apply a framework of criteria for
selection or rejection of texts, depending on the purpose of the corpus; for
example, gathering domain-specific information; terminology building;
identifying text-type conventions; identifying relevant language structures
— syntax, collocations, synonyms; even comparing layout and other
typographical conventions.

Documentary research for information should also be accompanied in
parallel by textual analysis of text typological features (Trosborg, 1997;
Chesterman, 1997; Nord, 1997), to ensure production of a functionally
adequate and acceptable TL text. We support Snell-Hornby's claim (1992:
17) that effective degree programmes in translation should expose students
to a wide range of subject areas and text types, to give experience in
documentary research, knowledge acquisition and awareness of text
typological conventions for the different text types. Thus the student is
trained in processes and in the practice of applying these processes. A

programme may wish to select texts from the most commonly required genres; these may include, *inter alia*, business, general scientific, journalistic, sporting, entertainment industry, and travel.

A further advantage of using corpora, in addition to increased subject knowledge and analysis of text-type conventions, is the scope for increasing language competence. Instead of correcting a student's work either by indicating the nature of the error or writing a correct version above the error (the latter is not to be recommended for obvious reasons), the tutor can ask the student to conduct a search of his/her corpus, to identify examples of a similar type in the corpus and to propose an explanation as to why his/her choice was less appropriate in that particular context. To do this, the student must first have access to a corpus on a topic, which contains texts of a similar type and communicative function, to ensure comparability of register, addressee, use of language etc. Not many such corpora are available, but with the Internet it is not difficult to compile a corpus, provided clear criteria for selection and inclusion of texts have been agreed in advance. We are currently working on a project funded by the Higher Education Funding Council for England, to look into how using corpora can assist advanced learners in becoming more proactive in the autonomous development of their language competence.

Translators Tools — Conducting Searches

To conduct searches, one commonly used tool is *Wordsmith* (others include *MonoConc* and *Word Cruncher*), which allows a search of a monolingual corpus, to look for Key Words in Context — a KWIC search.

Using Wordsmith

"*Wordsmith* is a concordancer which finds and displays, in easy to read format — KWIC (key words in context) — all occurrences of a search term and minor variations thereof." It works on a corpus of texts that have been collected in electronic form. Bowker (1998) describes using texts gathered from a commercially-available series of CD-ROMS, *Computer Select*, which contain English language articles from hundreds of publications dealing with computer-related topics. Careful searching on the Internet can discover other such sources for different languages.

Search strings can be determined for number of words, to look at use in a sentence, use in collocation, frequency of use, use with another specified term, variation of the same lemma; also to give statistical information about the importance of use of a language pattern, and to differentiate between idiolect and sociolect. It is useful in the detection of comprehension errors of a source message and production errors, including incorrect or inappropriate choices of syntax or lexis, register and idiom, in a target text.

Such searches have several advantages over dictionaries: selection of texts means that the most up-to-date data will be available, for a specified domain of use and type of addressee; information on phraseology, collocation and use in context will be more extensive; and above all, examples will be authentic examples of current use, not ones constructed artificially according to prescriptive predictions.

In comparison with using traditional grammar books, *Wordsmith* requires students to exercise greater analytical skills, to draw on and develop their intralingual competence not only in the target language but also in the mother tongue. With respect to domain-specific knowledge, it can also help students to develop their knowledge of key concepts in a subject field by looking at use in context of the relevant terms. It can further help in developing comparative and contrastive awareness of use of syntax and lexis, by comparing results of searches on two or more monolingual comparable corpora.

Experiments would tend to suggest that students can learn more from this kind of research than from a sterile and often complex, hence inaccessible, grammar book of the traditional kind. This is particularly true in the United Kingdom, where traditional teaching of grammar has been sacrificed over the past decades on the altar of a "communicative" approach to teaching language, which has resulted in a generation of trainee translators with little formal training in how language systems work. The obvious corollary to this approach to language development is that it empowers students to discuss alternatives with the tutor, to disagree and to express preferences. Since one of the aims of a translation programme is to enable students to make choices on the basis of sound reasoning and selection from possible alternatives, this is an outcome to be desired.

It is also helpful for the trainee translator to have the opportunity to search for examples of translation choices made by other translators — again, to be able to refer to authentic examples in context, within a given domain, to weigh these up and to assess the relevance of the strategies thus found, in relation to the task in hand. One tool which offers this facility is *Multiconcord*.

Using *Multiconcord*

Multiconcord is a parallel concordancing software, developed within an European Union Lingua initiative under the leadership of the University of Nancy, France. It operates on a corpus of parallel, translated text pairs which have previously been marked up. The corpus is designed to conform to the Text Encoding Initiative (TEI), using Standard Generalised Markup Language (SGML).

The software will search through a pre-marked corpus. It is a tool which allows the translator to specify a key word or phrase search/search string in one language. *Multiconcord* will find both the citations for that string in the source language and the corresponding sentences in the target language. Bilingual examples can be displayed horizontally or vertically, in a sentence or in a paragraph. In this way, trainees can identify, for example, similarities or differences in collocative use; SL polysemy and how this is resolved through TL synonyms; different ways of dealing with pragmatic problems for translation; ideas for overcoming text specific problems.

Multiconcord allows tutors to create tests to help trainee translators and students of language to develop their language competence and check their answers. It runs on a Windows environment and the designers are currently extending the languages offered, based on demand.

The minimal amount of encoding keeps the corpus at a manageable size for storage and also makes it easier for users to mark up their own texts. Texts included in the corpus have to be aligned by paragraphs in a one-to-one relationship. Results can be displayed at the level of the sentence or the paragraph. The constraint of paragraph alignment does, of course, restrict the type of texts that can be used with this tool, or else will involve realignment of texts prior to encoding. Further details of this tool can be found in, for example, King (1997) and King and Woolls (1996).

Corpus Studies and Text Typological Conventions

One type of translation problem (Nord, 1991, 1997) for which *Multi-concord* cannot really offer insights is the area of text typological conventions. The requirements of marking up texts to be searched by *Multiconcord* mean that texts have to be aligned at the level of paragraphs, also that even within paragraphs, there have to be corresponding sentences, otherwise sentences cannot be matched. This is why the texts of the European Union are particularly suited to this kind of search, as the demands of this multilingual environment require that all texts be structured according to the same sequence of information and ideas, even down to the division and layout of paragraphs, for comparability across different language versions. This is also why *Multiconcord* was produced initially with an accompanying corpus of European Union parallel texts in different languages. However, students can be trained to select and mark up their own texts to compile a new corpus to be searched by *Multiconcord*. This approach offers a blend of qualitative and quantitative searching, depending on how many texts are searched.

However, for students to develop greater awareness of text type conventions, the most effective approach is the qualitative one, involving detailed analysis of a corpus of texts of a given type, topic, length and focus. This is the real nitty-gritty of analysis, no computer tools to save time, just hard work and repeated study. It is, however, not difficult to compile a small corpus of some 25-30 texts in a given domain. Again, providing the selection criteria are carefully constructed to ensure both comparability and representativeness of the texts chosen, the corpus can yield some fairly useful initial data. For trainee translators this methodology offers both training in the compilation of a reliable corpus and experience in hands-on analytical techniques, leading to the intended outcome of text typological awareness. Even if the corpus is not totally balanced in terms of correspondence to key criteria such as text size, type, date of publication, author, etc., it can still offer useful and interesting data for analysis. Needless to say, storing texts in electronic form makes it easier to identify and match paired text segments according to a descriptive translation studies approach, then to correlated findings into logical conclusions.

All of these uses of domain-specific corpora, monolingual or bi-and multi-lingual, are examples of what Tim Johns describes as "data-driven learning." Students learn from the real world, from a descriptive and comparative approach as advocated by such translation scholars as Toury (1995). The final step in this process would then be to ask students to compare their findings with the examples proposed/prescribed by the traditional grammar book, to note differences in contemporary use, to speculate why and how this has happened (dynamics of language evolution; influences from other sources — text types, languages, domains; the role of cultural evolution, etc.). If students are asked to keep a translation diary (Beeby, 2000) of all the stages involved in this process, they will quickly see a patterned search method, a way of organising and interpreting findings in relation to their own language competence and a reliable means of justifying their own choices. This is all thorough training in how to survive in the professional environment.

Compiling a Corpus

We have noted the need for initial determination of criteria for selection and inclusion in a corpus. These criteria will vary in accordance with the intended use of a corpus. For highly-specialised, narrow focus studies, the criteria will be concomitantly more exhaustive and exclusive in nature. For the purposes of training, however, it may be more useful to construct corpora which can be used for variety of purposes, so that the criteria for selection may be less exhaustive, for example, basic criteria would relate to domain of activity, text type, length, time of production communicative function and purpose. (Reiss, 1977; Nord, 1997; Vermeer, 1996) In relation to date of publication, relative synchronicity is advisable, within a range that will be dictated by the domain, depending on the rate of evolution of knowledge and use of terminology, unless the specific aim is to study evolution of use over time.

Texts can be collected in printed form as well as from electronic sources, with the obvious disadvantage that these will have to be converted, by scanning or retyping, into electronic form for use by search tools such as *Wordsmith* and *Multiconcord*. Printed texts are, however, eminently suitable for qualitative analysis for text type conventions and if the corpus is small, as is often the case when students have to compile their own corpus, it can also serve for searches of a linguistic nature and for subject knowledge.

Texts in electronic form can be found on the Internet, through key word searches, from newspaper archives and other online databases. They can also be obtained from CDs, from company sources such as *Computer Select* (see page 143), from a company's in-house archives and from other translators. For students, their own work can be saved as a corpus for reference and analysis, provided the texts entered are correct in all aspects (language, content, layout, conventions etc). However, the main source for electronic data has become the Internet, particularly now that publishers are exploring the option of online publishing.

Using the Internet as a Translation Resource

It is arguably reliable, or almost, to say that the Internet has become the most widely-used resource for students at all levels, for a variety of purposes: to find key papers and articles to complement printed material resources; to search for elusive ideas, up-to-date information and terms; to confirm hunches or support hypotheses — the list is never-ending.

The translator of the new millennium cannot afford to ignore the potential of this vast resource and so it would seem essential that those responsible for training the translators of the future, within an academic environment which seeks to provide a solid foundation in both theory and practice, should fully address the complexities of this resource and its implications for training and practice. Included in this training should be an awareness of the limitations of such internet sources in terms of reliability, relevance, quality of the work and being up to date.

The Internet as a Vast Mono- or Multilingual Corpus

For the translator, there is a bewildering array of websites offering links to translation resources (dictionaries, glossaries, papers and lecture-notes, agencies, individual profiles, discussion-lists and many others). These are, on the whole, additional tools which offer speed of access and up-to-date references. However, the Internet offers far more than these simple tools and sources. It is, in effect, the largest corpus of material ever known, a heterogeneous collection of authentic material in its crudest form, just waiting to be discovered and used purposefully.

The Internet as a Synthesis of Other Translation Resources

Of course we train students to use other electronic resources, such as online dictionaries, specialist databases and free machine translation software, CD-ROM and computer software programmes, for all types of translation task, as well as for the compilation of a corpus.

It is essential that students not only seek to use electronic resources but that they demand the opportunity to use these, both during university studies and in the professional environment, in order to do justice to their own potential and to whatever task they may be set. The Internet does not replace traditional resources but complements them. It also gives access to many of the traditional resources online, so that from a translation work station the translator can access, via the computer, a full-range of types of sources. We need to train translators not only to use every possible source but to do in full critical awareness of the relevance and possible limitations of each source.

Conclusion

We have shown how, at Aston University, we develop and exploit the use of the Internet as a vast resource of parallel texts (in terms of monolingual, bilingual and/or multilingual comparable and/or translated parallel corpora) through the different levels of the programmes. Corpora form an integral part of the learning and assessment process in these programmes. Use of corpora is situated within a focus on criteria for preliminary source text analysis, for target text (TT) production and TT evaluation/quality control.

This leads also to a consideration of how the global corpus of translated texts, or texts produced in multilingual versions, affects the use of language in translation and requires us to review our views of fundamental concepts of translation theory, such as the changing nature of culture-specificity — who is the new target audience? Also, we need to consider the role and evolution of text-type conventions in an increasingly multinational situation of text production through translation. A further question that can be addressed through use of electronic corpora is the increasing cross-cultural contamination of text-type norms and use of language, particularly in domain-specific situations. A forthcoming study (Schäffner and Adab, 2001) will address this question of hybridity within the European Union, however the scope is wide within the multiplicity of international translation environments. For example, Professor Chan Tak-hung, of Lingnan University, published a paper (2000), on the internet website of the e-mail discussion group, *Translat 2000*, in which he describes "imports from the West" into translation studies in China:

> Global capitalism has infiltrated China's cultural landscape not only with its commercial mass-culture products but also with its academic, intellectual products, namely contemporary Western "theory." (Liu, 1996)

He discusses how concepts and approaches from Western translation theory have found a place in translation studies and university translation programmes in China, although he notes that these theories have yet to assert a strongly-felt presence in applied translation performance, as "operative" guidelines for a new kind of translation practice. This may already be happening: when it becomes sufficiently widespread, the demands of Western theories such as functionalism will most probably require different types of translation strategy to deal with new text types. This is when you will begin to notice the effects of hybridity, the mingling of traditions and features from more than one culture in the production of a functionally adequate text. This is also when students will need to take into account how best to combine the theoretical concepts offered by different traditions, to produce the best results — in other words, to adopt a truly functional approach to each individual task of translation.

We have tried to show what our aims are and how we try to ensure that students achieve these. It is our view that the final aims of a translation programme are to help students develop:

- *a.* the ability to produce a functionally adequate translated text for any translation situation;
- *b.* all the different competences a translator needs to achieve this goal;
- *c.* sensitivity to the needs of a situation;
- *d.* sensitivity to changes in use of language;
- *e.* sensitivity to the changing demands of the professional translation environment.

We could not achieve all the aims of our programme in terms of developing translation competence and providing a grounding in the key demands of the professional translation environment without access to all of these computer-based resources. However, of all of these, we would argue that the principles of the use of corpora in translation are fundamental to this training and it is an irrefutable fact that properly used, the Internet can provide access to a range of texts which can serve in different ways as corpora for the purpose of translation.

References

Anderman, Gunilla and Margaret Rogers (2000). "Translator Training between Academia and Profession — A European Perspective," in Christina Schäffner and Beverly Adab, eds., *Developing Translation Competence*. Amsterdam and Philadelphia: John Benjamins Publishing Company, pp. 63–76.

Beeby, Allison (2000). "Evaluating the Development of Translation Competence," in

Christina Schäffner and Beverly Adab, eds., *Developing Translation Competence*. Amsterdam and Philadelphia: John Benjamins Publishing Company, pp. 185–98.

Bowker, Lynne (1998). "Using Specialised Monolingual Native Language Corpora as a Translation Resource: A Pilot Study." *Meta*, Vol. 18, No. 4, pp. 631–51.

Catford, John C. (1965, 1974). *A Linguistic Theory of Translation: An Essay in Applied Linguistics*. London: Oxford University Press.

Chan, Tak-hung (2000). "Translation Studies in China: The Impact of 'New' Translation Theories." http://www.translat2000.com/discussion/newtheory.htm.

Chesterman, Andrew (2000). "Teaching Strategies for Emancipatory Translation," in Christina Schäffner and Beverly Adab, eds., *Developing Translation Competence*. Amsterdam and Philadelphia: John Benjamins Publishing Company, pp. 77–90.

King, P. (1997). "Parallel Corpora for Translator Training," in *Proceedings of "Translation and Meaning."* Lodz, pp. 393–402.

King, P. and D. Woolls (1996). "Creating and Using a Multilingual Parallel Concordancer," in Marcel Thelen and Barbara Lewandowska-Tomasczyk, eds., *Proceedings of "Translation and Meaning: Part 4,"* pp. 459–66.

Kussmaul, Paul (1991). "Creativity in the Translation Process: Empirical Approaches," in Kitty M. van Leuven-Zwart and Ton Naaijkens, eds., *Translation Studies: The State of the Art: Proceedings of the First James S. Holmes Symposium on Translation Studies*. Amsterdam: Rodopi, pp. 91–101.

Kussmaul, Paul (1995). *Training the Translator*. Amsterdam and Philadelphia: John Benjamins Publishing Company.

Li, Defeng (2000). "Tailoring Translation Programmes to Social Needs: A Survey of Professional Translators." *Target*, Vol. 12. No. 1, pp. 127–49.

Liu, Kang (1996). "Is There an Alternative to (Capitalist) Globalisation?" *Boundary*, Vol. 2, No. 23, p. 210, quoted in Chan Tak-hung (2000), "Translation Studies in China: The Impact of 'New' Translation Theories." (http://www.translat2000.com/discussion/newtheory.htm).

Neubert, Albrecht (2000). "Competence in Language, in Languages, and in Translation," in Christina Schäffner and Beverly Adab, eds., *Developing Translation Competence*. Amsterdam and Philadelphia: John Benjamins Publishing Company, pp. 3–18.

Neubert, Albrecht and Gregory M. Shreve (1992). *Translation as Text*. London: Kent State University Press.

Newmark, Peter (1990). *A Textbook of Translation*. New York: Prentice Hall.

Newmark, Peter (1988). *Approaches to Translation*. Oxford and New York: Pergamon.

Nord, Christiane (2000). "On-line Translation Symposium." http://www.fut.es/~ãpym/symp.html.

Nord, Christiane (1991). *Text Analysis in Translation: Theory, Methodology and*

Didactic Application of a Model for Translation. Amsterdam: Amsterdamer Publikationen zur Sprache und Literatur.

Nord, Christiane (1997). *Translating as a Purposeful Activity: Functionalist Approaches Explained*. Manchester: St Jerome Publishing.

O'Hagan, Minako (1996). *The Coming Industry of Teletranslation*. Clevedon: Multilingual Matters Ltd.

Reiss, Katharina (1977). "Texttypen, Übersetzungstypen und die Beurteilung von *Übersetzungen*." *Lebende Sprachen*, pp. 97–100. Published in English in Andrew Chesterman, ed. (1989) and Christiane Nord (1991, 1997).

Schäffner, Christina and Beverly Adab, eds. (2000). *Developing Translation Competence*. Amsterdam and Philadelphia: John Benjamins Publishing Company.

Schäffner, Christina and Beverly Adab, eds. (2001). *The Hybrid Text in Translation*. Special edition of: *Across Languages and Cultures* (forthcoming).

Snell-Hornby, Mary (1992). "The Professional Translator of Tomorrow: Language Specialist or All-round Expert?" in Cay Dollerup and Anne Loddegaard, eds., *Teaching Translating and Interpreting: Training Talent and Experience*. Amsterdam and Philadelphia: John Benjamins Publishing Company.

Tirkonnen-Condit, Sonja (1991). *Empirical Research in Translation and Intercultural Studies*. Tübingen: Gunter Narr Verlag.

Tirkonnen-Condit, Sonja (1996). "What Is in the Black Box ? Professionality in Translation Decisions in the Light of TAP Research," in V. Lauer, *et al.*, eds., *Ubersetzungswisenschaft in Ümbruch. Festschrift für Wolfram Wilss zum 70 Geburtstag*. Tübingen: Gunter Narr Verlag.

Toury, Gideon (1995). *Descriptive Translation Studies and Beyond*. Amsterdam and Philadelphia: John Benjamins Publishing Company.

Toury, Gideon (1980). *In Search of a Theory of Translation*. Tel Aviv: The Porter Institute.

Trosborg, Anna, ed. (1997). *Text Typology and Translation*. Amsterdam and Philadelphia: John Benjamins Publishing Company.

Vermeer, Hans J. (1996). *A Skopos Theory of Translation: Some Arguments For and Against*. Heidelberg: TextConText.

Vermeer, Hans J. (1986). *Voraussetzungen für eine Translationstheorie*. Heidelberg: TextConText.

Vinay, Jean-Paul and Jean Darbelnet (1977). *Stylistique Comparée de l'Anglais et du Français*. Paris: Didier.

Other Resources

1. http://europa.eu.int/eur-lex/en/lif/dat/1999/en_399Y0317_01.html.
2. *Computer Select*, series of CD-ROMS, from Ziff-Davis Publishing, Computer Library, New York, updated monthly and published in December each year.

3. For *Multiconcord*: contact David Woolls at: 100343.2362@compuserve.com.
4. For details of Tim Johns' work and of data driven learning sources, see: http://web.bham.ac.uk/johnstf.
5. For *Wordsmith*: http://www1.oup.co.uk/elt/catalogue/Multimedia/WordSmith Tools3.0/.

Computer Technology and Translation — Friends or Foes?

Carrie Chau Kam Hung and Irene Ip Kwok Chun
Division of Language Studies
City University of Hong Kong

"The integration of technology into higher education is creating opportunities for educators to think in entirely new ways about how to define the focus of study ..." (Gates, 2000:270) This idea of integrating technology into higher education has been put forward by Bill Gates, the founder of Microsoft. Indeed, he upholds the positive role of computer technology and encourages its application in higher education as it can "support learning at all levels in an exhilarating prospect" and lead to "benefits of improving education." (2000:269)

In this Information Age, computer technology has been strongly recommended and widely accepted by educators and also people in other fields. As suggested by Halal and Leibowitz (2000:82), "the key technology in future education is interactive multi-media" which "combines computer hardware, software, and peripheral equipment to provide a rich mixture of text, graphics, sound, animation" and so on.

Our previous studies on Computer-assisted Language Learning (CALL) focusing on the area of translation have also confirmed the positive role of computer technology in learning. However, there are also sceptical criticisms on the limitations of using computers in education. This paper attempts to explore further into the advantages and disadvantages of computer technology. Then, we shall report on a study that we have conducted in order to assess whether computer technology facilitates or hinders the learning of translation. The discussion will be directed particularly towards the application of computer technology in the field of translation with a view to ascertaining the role of computer technology in relation to translation — whether it helps to facilitate translation and thus is a friend, or it may hinder translation and so becomes a foe.

The advantages of computer technology are mainly related to its flexibility and interactivity. According to the supporters of computer technology in education, computers let users pursue their interest as far as they like. Students can "receive training when and where they need it." Computers allow the flexibility for students "to stop at any point (in the lesson) and come back later." (Halal and Leibowitz, 2000:83) Moreover, as computers offer interaction, they allow students to have self-administered quizzes. In other words, the students will have more control over their learning. Indeed, computer technology may be viewed as an important tool that can improve education. Other advantages of using computers for language learning include the interesting and stimulating approaches to learning and also the engaging activities they can provide.

As for the arguments against computers in education, the main criticism concerns the "sleepwalking attitudes" of many people towards computers. This refers to the unquestioned acceptance of computer technology and even the attempt to deify it only because it is the trend without understanding clearly what it is all about. As firmly expressed by Postman (2000:282), "I am arguing against our sleepwalking attitudes toward it, against allowing it to distract us from more important things, against making a god of it." Another disadvantage of the computer is its lack of interaction with others as students mainly learn in isolation with a machine, and hence may lead to dehumanisation. Moreover, some people are afraid that computers will dominate the learning experience. Instead of solving problems, computers may worsen them. Some sceptics, like Postman, even suggest that it is not a new technology, but it requires "a new species" of students (and teachers) instead. Indeed, it does require some basic knowledge of computer technology before the students and teachers can make use of it for learning or teaching. It has even been pointed out by Postman (2000:281), that "what we need to know about technologies is not how to use them but how they use us."

The above-mentioned advantages and disadvantages are mainly on the use of computers in education, but how do they apply to the field of translation? In order to explore further into the advantages and disadvantages of applying the computer in translation, we conducted a study with a view to comparing the different modes of learning — the use of computer versus the traditional method of paper and pen.

Design

Since vocabulary acquisition is an essential component in translation, we would like to adopt this as the content of our study. However, it is hard to find ready-made computerised learning programmes on specific translation training tasks; we have specially designed a package for learning bilingual (Chinese-English) vocabulary on social issues and mass media as the manipulated variable in our study. Given the learning task is the same but the learning method is different, it can be assumed that the performance of the participants using the two methods may be an indicator of the effectiveness of the method adopted.

The design of the study is in the form of within-group matched-subjects and between-group comparison. In order to formulate the computer package, we input the collected bilingual vocabulary concerned from newspapers and magazines. The nature of the vocabulary can be generally divided into three categories: general, mass media and government structure. There are 20 items in each category, and so there is a total of 60 items of bilingual vocabulary on social issues and mass media. As for the subjects, there were two main groups, the experimental group using computer as the learning mode and the control group adopting the traditional method of paper and pen as the learning mode; both were made up of seven Year 3 and seven Year 1 students of the Translation and Interpretation Stream, Division of Language Studies, City University of Hong Kong. They were recruited on a voluntary basis.

The design was based on the following hypotheses:

(1) The Year 3 Translation and Interpretation (TI) students would perform better than the Year 1 TI students;
(2) Both Year 3 TI students and Year 1 TI students would perform better in Test 2 than in Test 1;
(3) The students would prefer using the computer to the traditional method for learning the vocabulary;
(4) The students in the experimental group would have greater improvement than those in the control group.

Procedure

In order to determine the background knowledge of the target vocabulary, the subjects were given a test (Test 1). The experimental group was asked to use the computer for the learning task while the control group used the printed material and writing instead. During the course of the test, one-third (20 items) of the input vocabulary list came out randomly. The subjects had 60 seconds to translate each of the given items (shown one at a time) on the screen. Of the 20 items, 10 were to be translated from English into Chinese and the other 10 vice-versa. When they responded in Chinese, the subjects could write on the writing tablet. When they answered in English, the subjects could input the information directly through the keyboard. The individual results could be printed straight from the computer, so the subjects knew about their performance immediately from the scores.

Then, they took part in a 20-minute workshop in the form of a computer game to reinforce their understanding of the vocabulary. The game-workshop involves interactive learning, drills and practice through the computer. During the session, the previously input vocabulary would be generated randomly in Chinese or English for the participant to translate. In each attempt there are seven chances (indicated in the form of seven smiling faces) for the participant to make guesses. The wrong guesses would be shown as reminders for the participant to avoid them. After one to two week(s), the subjects came again for the second test (Test 2).

Similar arrangements were made for the control group. However, instead of showing the vocabulary list on the computer screen, the subjects were given a written list and had to finish Test 1 within 15 minutes by writing down the answers on the paper. Then they were given 20 minutes to get familiarised with the vocabulary (familiarisation exercise) to replace the game-like workshop. This was what they usually did with the traditional method and did not involve the computer at all. After one to two week(s), the subjects came again for the second test (Test 2).

The purpose of the above arrangements is to compare the results of the two tests. We would like to check if the results of Test 2 could be better after Test 1 and also the game-workshop/familiarisation exercise. At the end of Test 2, all the students were asked to fill in a question-naire.

Findings

Generally speaking, all of the hypotheses were supported, that is to say, for the between-year comparison the Year 3 TI students (mean percentage of accuracy is 50.70% in Test 1 and 73.20% in Test 2 respectively) performed better than the Year 1 TI students (mean percentage of accuracy is 41.78% in Test 1 and 64.65% in Test 2) and both groups did better in Test 2 than in Test 1 for the between-test comparison (pre- and post-tests). Most subjects (nearly 70%) preferred the computer to the printed list for learning the vocabulary. As for the between-medium comparison, if we focus on the improvement rate, then the subjects in the experimental group (i.e. those who took computer as the medium) had a much higher degree of improvement. The results are indicated more clearly in Table 1 as follows:

Table 1. Mean Percentage of Accuracy in Tests and Improvement Rate

	Test 1	Test 2	Improvement Rate*
Experimental Group			
Year 3 Students	40.70%	65.70%	61.43%
Year 1 Students	32.85%	57.15%	73.97%
Control Group			
Year 3 Students	60.70%	80.70%	32.95%
Year 1 Students	50.70%	72.15%	42.31%

* Note: $\dfrac{\text{Score of Test 2} - \text{Score of Test 1}}{\text{Score of Test 1}} \times 100\%$

Between-year Comparison

Although the subjects were all majoring in Translation and Interpretation at City University of Hong Kong, and they responded to an invitation on a voluntary basis, they were quite different both in terms of training and experience as well as in their mind-set. The Year 1 students only had initial training and basic knowledge of translation. Therefore, translation tasks for them were challenging and they did not yet have the urgency of becoming professional translators. As for Year 3 students, they had two more years of training and knowledge. So, they were more familiar with translation tasks. Since they were about to graduate, they had the urgency of finding jobs or furthering their studies in translation or related fields. It is understandable that these batches of students had quite significant difference in their performance in the tests and also responses to the questionnaire.

Between-test Comparison

Based on the within-group matched-subject design for comparison of pre- and post- test results, both the subjects in the experimental group and control group had improved in their Test 2 scores. This also implies that both Year 3 and Year 1 TI students had made improvement. It is reasonable that after the exposure to the target list of vocabulary in Test 1 and also in the workshop/exercise, the subjects had more ideas about the related bilingual vocabulary.

Between-medium Comparison

The results in Test 1 reflect that the subjects in the control group (writing method) had better background knowledge of the target vocabulary (mean percentage of accuracy is 55.70%) than those in the experimental group (mean percentage of accuracy is 36.78%). Both groups performed better in Test 2. But if we focus on the improvement rate (see Table 1), then it is obvious that those in the experimental group had a higher degree of improvement (28.48% for Year 3 students and 31.66% for Year 1 students) than those in the control group, thus verifying that the computer did play a significant role in this study in facilitating learning of bilingual vocabulary.

Responses to the Questionnaire

The evaluation questionnaire results are encouraging. The students were asked to fill in a questionnaire at the end of Test 2 to rate the four dimensions (*difficulty, interest, effectiveness and duration*) on a 10-point scale with 1 as *very easy, very interesting, very effective* and *very short* respectively. Most of them expressed that the difficulty of the activity was appropriate. For the experimental group, the average score was 5.07 and for the control group, 5.36. As regards the point of interest, the experimental group found it interesting (with an average score of 2.93) whereas the control group found it less interesting (with an average score of 5.50). In terms of effectiveness, the experimental group considered the activity to be more effective as compared with the control group. The average scores were 3.21 (more effective) and 4.64 (less effective) respectively. In terms of duration, the scores of the two groups are quite close (4.93 for the experimental group and 5.07 for the control group), indicating that both groups found the duration to be acceptable.

Nearly all of the subjects showed interest in participating in similar activities later. The majority (over 80%) of them found that the related vocabulary list was useful to translation and interpretation students. Over 90% of the subjects felt that the vocabulary would be useful to their future career in the translation/interpretation field. About 90% recommended introducing similar activities in lessons. Of the three categories of the vocabulary list, the government structure was considered to be the most useful, followed by mass media. This may be related to the change of the government structure after the Handover of Hong Kong back to China.

Conclusion

From the above findings, it can be concluded that all of our hypotheses have been generally supported. The use of computer can help students learn the bilingual vocabulary list effectively. This is especially obvious with Year 1 students in the experimental group whose improvement rate is 73.97%. It may imply that Year 1 students may benefit more from using computers for translation learning tasks. It is possible that they are more ready to learn new things through the computer. The fact that the participating students indicate that they would join similar activities in the future may reflect their interest in using computers for learning.

The present study has adopted two most effective vocabulary methods — vocabulary lists and vocabulary tests. Many subjects have developed or will develop the habit of compiling glossaries, and this reflects the function of vocabulary lists. Since the subjects can find out their own results from the tests, including the correct and incorrect answers, these tests could enable them to have a deeper impression of the related vocabulary. As pointed out by Fenrich (1997:204), "feedback helps learners improve their subsequent actions and responses ..."

As Paulston (1992:167) points out, we need to look to "socio-economic and cultural factors" related to language acquisition. This is exactly what the content of the vocabulary in the study reflects. Through the study, the validity of computer-assisted language learning has once again been confirmed. Through the interaction with the computer, participating students have more active control during the learning process and they have again verified the value of this new learning mode.

Indeed, the present study has also attested to the other advantages of the computer apart from giving more control to students over learning. The flexibility of the random test and practice items from the vocabulary list may ensure more that learning would really take place. The interactivity with the computer in both the tests and the game-workshop help to motivate students to learn more. The capability of the computer in generating self-administered tests would enable students to take a more active role in learning. Moreover, the engaging activity in using the computer would enhance concentration on the part of the users and reduce the unnecessary distractions to a certain extent.

In order to eliminate the disadvantages of the computer, we need to be aware of them and see the use of the computer in perspective. It is not something to be deified or accepted without questions. Rather, it is a tool that can escalate our learning if used properly. Although it is only a machine, the users need not be isolated or dehumanised. On the one hand, the computer can help us link up to others and contact them even across the miles. On the other, we can still interact with others who are co-learners in front of the machine through discussion and sharing of the skills in mastering the latest development of computer technology. Hence, the computer would not dominate our learning experience, but would help to add more meaning to it instead. All along we need to remember that the computer may lead to unexpected technical problems despite its advanced development. But we should try to face them positively and find out the feasible solutions; merely avoiding them or becoming so threatened by them as to discard the use of the computer altogether would be equally inappropriate.

As for the future direction of study, it is suggested that while we should consolidate and expand the existing list, we should also branch out into other areas like business, law, information technology and slang. As for the learning process, some subjects prefer to have an option of either using the writing tablet or keying in the Chinese characters on the keyboard. Another suggestion is to have a voice over the text so that the subjects can also learn the pronunciation of the related vocabulary. The design of the game could be more varied with better visual and sound effects. If possible, more subjects may be recruited for future studies.

Overall speaking, we may conclude that computers may be a foe of translation owing to possibly unexpected technical problems and also if we treat it inappropriately, but computer technology can certainly be a friend if we can capitalise on its constructive potentials and eliminate its destructive risks.

References

Fenrich, Peter (1997). *Practical Guidelines for Creating Instructional Multimedia Applications*. Fort Worth: The Dryden Press.

Gates, Bill (2000). "Linked Up for Learning," in William E. Vesterman and Josh Ozersky, eds., *Readings for the 21st Century: Tomorrow's Issues for Today's Students* (4th ed.). Boston: Allyn and Bacon, pp. 269–76.

Halal, William E. and Jay Leibowitz (2000). "Telelearning: The Multimedia Revolution in Education," in William E. Vesterman and Josh Ozersky, eds., *Readings for the 21st Century: Tomorrow's Issues for Today's Students* (4th ed.). Boston: Allyn and Bacon, pp. 82–90.

Paulston, Christina Bratt (1992). *Sociolinguistic Perspectives on Bilingual Education*. Clevedon: Multilingual Matters Ltd.

Postman, Neil (2000). "From the End of Education," in William E. Vesterman and Josh Ozersky, eds., *Readings for the 21st Century: Tomorrow's Issues for Today's Students* (4th ed.). Boston: Allyn and Bacon, pp. 277–84.

Information Technology in Translator Training: Reflections on an Aborted Comprehensibility Test of Machine-translated Texts

Li Defeng
Department of Translation
The Chinese University of Hong Kong

Vilson J. Leffa (1994) conducted a study on the comprehensibility of machine-translated passages in which secondary school students were requested to read two different versions of the same passage, one a machine translation and the other a human translation, from the English original into Portuguese. From this, he concluded:

> The results of the study ... showed that readers, in general terms, can read and understand machine-translated passages with the same level of proficiency as they read passages translated by professionals. The grammatical errors found in machine translations did not significantly affect the subjects' comprehension scores. This suggests that MT, although unable to produce error-free translations, can be used for comprehension purposes. (Leffa, 1994: 399)

Inspired by Leffa's findings, I decided to study the comprehensibility of texts generated by some well-known English-Chinese translation software programmes. My purpose was to determine whether these programmes can truly fulfill the purpose, as many of them claim, of providing people with a general level of reading comprehension. For instance, a report on *TransEasy* stated that, "In the open test of Chinese-English machine translation systems held by the Steering Committee of the Chinese Hi-Tech R&D Plan in March 1998, this system got a good result, about 70% of the translation is understandable." (Liu and Yu, 1998:

516) Aware of the unsatisfactory performance of Chinese-English translation software, I decided to limit the study to English-Chinese translation software only.

Similar in design to Leffa's study, I used a journalistic text of about 650 words, which I took from the American Broadcasting Corporation's homepage. I ran it through several well-known English-Chinese translation software programmes, namely, *Dr. Eye*, *Oriental Express*, and *Huajian*. Unfortunately, the results were truly discouraging. In addition to such funny mistakes as translating former US President Bill Clinton's name as "賬單克林頓" (both *Dr. Eye* and *Oriental Express*), and phrases such as "where he once again pressed the flesh" into "再一次按肉" (*Oriental Express*), "再一次壓肉" (*Huajian*) and "再一壓 (下) 了這個按肉" (*Dr. Eye*), the entire output of machine-generated passages was absolutely incomprehensible, except for a few sentences which could be partially understood. By way of example, let's take a look at the following seven pairs of sentences.

Source Text	Target Text
(1) "Almost 200 years ago at the beginning of relations between the United States and Vietnam, our two nations made many attempts to negotiate a treaty of commerce," he said.	*"幾乎200年以前在開始時關係位於美國和越南，我們的二國家製作許多企圖商議商業的條約"他說。*
(2) Let us continue to help each other heal the wounds of war, not by forgetting the bravery shown and the tragedy suffered by all sides, but by embracing the spirit of reconciliation and the courage to build a better tomorrow.	*讓我們繼續幫助互相醫治戰爭的傷口，不由忘記勇敢顯示和悲劇遭受由所有邊，但是由包括reconciliation的精神和勇氣到建造好明天為我們的孩子。*
(3) Let me say emphatically, we do not seek to impose these ideals.	*讓我說強調地，我們不查找強迫這些理想的。*
(4) Only you can decide if you will continue to open your markets, open your society, and strengthen the rule of law.	*僅僅你能決定如果你願意那樣說的話繼續打開你的市場，打開你的社會和加強法律的規則。*
(5) Mr. Clinton said his visit was a	*Mr克林頓說他的訪問是美國善*

gesture of American goodwill towards Vietnam, that the U.S. had abandoned the animosity from the war and was now looking to the future.

(6) As he left the university auditorium he broke protocol by walking past security guards to a cheering crowd, where he once again pressed the flesh.

(7) The man grasped the President's hand and, with tears in his eyes, said: "I could never imagine a day like this."

意的姿勢向越南，那美國已放棄仇恨從戰爭和在現在照看未來。

當他離開大學禮堂他打碎協議田步行過去安全警衛鼓勵擠滿，在哪裡他再一次按肉。

那個人抓住主席的手和，含淚在他的眼睛，説："我能永不想像天喜歡這。"

In these seven sentences, generated by *Oriental Express*, only the underlined parts are somewhat intelligible. It is not possible to construct the general meaning of the text from simply reading the machine-translated text alone. The translation generated by *Huajian* seems no better. Except the following eight sentences, which are partially understandable, the rest of the translation is unintelligible.

Source Text	**Target Text**
(1) US President Bill Clinton received a standing ovation in the Vietnamese capital yesterday after acknowledging the pain of the Vietnam War and calling for a new era of friendship and co-operation.	*在承認越南War的痛苦并且要求一個友誼和合作的新的時代之后，美國比爾。克林頓總統昨天在越南語大寫字母方面收到一次經受的熱烈歡迎。*
(2) In his address at the University of Hanoi, broadcast live on Vietnamese television — the first such gesture of trust to be extended to a visiting head of state — Mr Clinton lamented the mistakes of the past and called on the young of both countries to learn the lessons of history.	*在在Hanoi的大學的他的地址，廣播靠越南電視生活——被給予一個訪問的國家首腦的第一個這樣的信任的手勢——克林頓先生悲傷過去的錯誤并且號召兩個國家的年輕學習歷史的課。*

(3) "Almost 200 years ago at the beginning of relations between the United States and Vietnam, our two nations made many attempts to negotiate a treaty of commerce," he said. "… [but] efforts failed because two distant cultures were talking past each other …"

"幾乎在在美國和越南之間的關系的開始200年以前，我們的兩個國家使很多嘗試談判一項商業的條約，"他說。"……除了〔的〕努力失敗，因為兩種遠的文化通過彼此是交談……

(4) "Let the days that we talk past each other be gone for good. Let us acknowledge our importance to one another. Let us continue to help each other heal the wounds of war, not by forgetting the bravery shown and the tragedy suffered by all sides, but by embracing the spirit of reconciliation and the courage to build better tomorrows for our children."

"讓我們通過彼此交談的那些日子永遠走開。讓我們為了相互承認我們的重要性。讓我們繼續互相幫助以忘記顯示的英勇治愈war的傷口并且悲劇以全部邊痛苦，但是給我們的孩子更好明天建設的以復交和勇氣的接受酒精。"

(5) "Let me say emphatically, we do not seek to impose these ideals … only you can decide how to weave individual liberties and human rights into the rich and strong fabric of Vietnamese national identity," he said. "Only you can decide if you will continue to open your markets, open your society and strengthen the rule of law …"

"讓我強調地說，我們不想辦法強加這些理想……"僅僅你能決定怎樣織個別的冒昧和人權進越南國家身份的富人和強壯的織品，"他說。"僅僅你一直能決定如果你一直將繼續開立你的銷路，開立你的社會并且加強法律性規則法治……"

(6) "In our experience, guaranteeing the right to religious worship and the right to political dissent does not threaten the stability of a society. Instead it builds people's confidence in the fairness of our institutions."

"在我們的經驗裏，保證對宗教的崇拜的權力和對政治不同意的正確不威脅一個社會的穩定性。代替它建造人們的對我們的機構的公正的信心。"

(7) He said the US people were changing their view of Vietnam. "America is coming to see your

他說美國人們正改變他們的越南的意見。"美國開始看出你國家，因為你已經要，作為

country, as you have asked, not as a war but as a country."

(8) Thousands raced on their scooters after the President as he crossed the city to the University of Hanoi, where they swamped the streets outside as he made his key speech.

(9) Mr Clinton is the first US president to visit the capital of a united Vietnam. His trip caps an eight-year period of reconciliation that started with Washington easing a trade embargo imposed after Hanoi's victory over the US-backed South in 1975.

(10) Mr Clinton said his visit was a gesture of American goodwill towards Vietnam, that the US had abandoned the animosity from the war and was now looking to the future. He again praised Vietnam for its compassion in helping to locate the remains of still-missing US servicemen and acknowledged "the staggering sacrifice of the Vietnamese people on both sides of that conflict."

(11) "The histories of our two nations are deeply intertwined in ways that are both a source of pain for generations that came before, and a source of promise for generations yet to come," he said. As he left the university auditorium he broke protocol by walking past security guards to a cheering crowd, where he once again pressed the flesh. "Thank you to you all," he said before moving towards an elderly man.

一war一war*但是作為一個國家。"*

千在總統之后在他們的小型摩托車上比賽，當他把城市錯過到Hanoi的大學，*他們在外面淹沒那些街道，當他做他的關鍵演說。*

克林頓先生是訪問一聯合的越南的首都的第一個美國總統。一個從緩和一次貿易禁止的華盛頓開始的復交 的八年時期強加的他的旅行帽子在Hanoi的對美國的勝利在1975年支持南方之后。

克林頓先生說他的訪問是對于越南的一個美國友好的手勢，美國已經從war放棄仇恨并且現在正注意將來。他再次稱贊越南它的在制止方面的同情設置仍然丟失的美國技術工的殘餘并且告知"蹣跚越南人們在那次衝突的兩側犧牲。"

"我們的兩個國家的歷史深被纏繞用以前來的代是兩個一個痛苦的源的方式，并且要來的一個代有前途的源，"他說。當他離開大學的大禮堂他以對一幹杯的人群經過護衛員走損壞議定書，這裏他再一次壓肉。"向你全部感謝你，"在移向一個漸老的人以前，他說。

Discouraging as the results of this trial run of machine translation software might be, I do not believe that the effort to develop English-Chinese translation software should be abandoned. On the contrary, more work should be done to strengthen such programmes. There are two reasons for continuing such efforts.

First, the enormous increase in the amount of translation that is required due to the development and subsequent globalisation of economic and information technology. Such a trend makes it imperative that technology be used to facilitate translation. On the global level, the worldwide market for translation and software and website localisation services is large and continues to grow. According to Allied Business Intelligence, this market was valued at $11 billion in 1999, and this number is expected to grow to $20 billion by 2004. (Sprung, 2000) A study by the European Commission even valued it at over $30 billion annually, with a growth rate of 15–18% per year. (Anobile, 2000) For English-Chinese translators and translation trainers, China's entry into the World Trade Organisation is particularly meaningful, as we can easily foresee a tremendous increase in the demand for English-Chinese translation.

In order to meet the growing needs of this market, more quality translators will need to be trained. However, training alone would not solve the problem entirely. Considering the indomitable amount of translation work and the increasing pressure for shorter turnaround time, technology-assisted translation needs to be improved and fully utilised to enhance efficiency. Especially with China's entry into the WTO, improving the existent English-Chinese translation software while devising new ones seems crucial.

A quick review of the history of machine translation and computer-assisted translation also reminds us that persistence and painstaking effort have brought about enormous progress over the past fifty years. During this time, machine translation has grown from a tantalising dream to a respectable and stable scientific linguistics enterprise with users, commercial systems, university researchers, and government participation. (Neubert, 1991; Farwell, Gerber and Hovy, 1998) So far, the European Commission is probably the largest and most successful user of machine translation services. Today, the professional translator's use of translation software to assist with translation is already a fact, which I have observed repeatedly when reading the e-mailed messages exchanged over the Internet between professional translators subscribing to several online translation discussion groups (e.g., *Translat 2000*, *Fanyi-L*). Fan (2000) reported his success in publishing three translated books using a computer-aided translation software programme, *Yaxin* CAT. All this shows that as long as we do not give up trying, we can only get better with our efforts to develop such programmes to assist us with our work.

Of course, as our efforts continue, the approach we adopt in studying machine translation will also be vital in bringing us closer towards our goal. In a recent article published in *The Linguist*, Wood criticized the practice of excluding professional translators and translation scholars from the research and development phase of computer translation. He contended that this was the major reason for the generally disconcerting situation of computer translation.

> Computational linguists, artificial intelligence researchers, language engineers, and (in their latest incarnation) cognitive neuroscientists have never really bothered to ask translators what it is they do [in the process of translation]. ... Most language engineers are monolingual, trapped in their own native tongue with a smattering of schoolboy Spanish or French ... (Wood, 2000:133)

Chan (2000) echoed Wood's argument:

> As we all know, the various machine systems for translating between Chinese and English, many of which are still far from satisfactory, have been mainly designed by those in the fields of computer science and linguistics, and rarely by scholars who have a good mastery of concepts in translation studies, in particular translation methodology for the rendition of modern scientific and procedural texts which are barely intelligible to the general reader. Secondly, machine translation works at the syntactical level, and this means that the methods frequently used by professional translators in translating sentences to produce effective sentences have not been put to good use in machine translation. (abstract)

Despite Chan's insightful proposal to include translators in the software development process, it would be unwise to get overly optimistic about the ability of machine translation. A good example of the dangers of being overzealous is former U.S. President Bill Clinton. Impressed with the marvels that technology produced, he stated in his 2000 State of the Union Address that "some researchers will bring about devices that can translate foreign languages as fast as we talk." It sparked anger among American translators, and was immediately challenged by translators in the U.S. and other parts of the world. Here is an excerpt from what former

American Translators Association (ATA) President Ann MacFarlane wrote to Mr. Clinton on behalf of the association.

> While technology has produced many marvels, machines that 'translate as fast as we can talk' are still a long way off. As you know from your work with interpreters in high-level meetings and negotiations, it takes experience, knowledge, native ability, and training to interpret foreign languages correctly. Despite the increasing compactness and cleverness of all the computing devices now on the market, human speech remains something that can be interpreted correctly only by human beings. (*ATA Chronicle*, March 2000:9)

As a result of the strong protest from professional translators and translation scholars, Mr. Clinton had to acknowledge in his reply:

> As we prepare for the opportunities and challenges of an increasingly globalized world, the need for dedicated competent translators will be even more important. (*ATA Chronicle*, September 2000:25)

In addition to continuing our efforts to improve English-Chinese translation software, the technological factor must be incorporated in translator-training curricula. If the use of technology does not in itself make a person a better translator, at the very least, it helps translators to research, process, and organise their work faster and more efficiently, and in formats that are in increasingly greater demand on the market. The uptake of information and communication technologies, and the integration of language engineering and technology within these environments, have transformed document management processing in commercial and public organisations. Industrial companies are using sophisticated software tools in all areas of document creation, terminology management, and translation. The Internet, in the span of just a few years, has revolutionised our attitudes with regard to information retrieval. The Internet provides access to a vast amount of information that was previously difficult or expensive to get.

It is obvious that innovation in language technology will play a crucial role in the future of all professional translators. However, changes in these commercial environments have not yet been fully reflected in the training of translators, who need to develop appropriate skills and knowledge in information and communication technology to satisfy the requirements of their prospective employers. The European Commission is among the first to see the importance and the necessity of a move in this direction. It recently supported a translator-training curriculum innovation project,

Language Engineering for Translation Curricula. (Maia, 1998) The project is designed to survey the curricula of translator-training programmes in several EC countries (including Germany, Spain, Portugal, Greece, Denmark, Belgium, and Luxemburg) and the needs formulated by translation professionals, associations, and clients. The EC will use its findings to establish a common basis for the elaboration and inclusion of language engineering components in B.Sc. and M.Sc. translator curricula in EC countries.

In comparison, translation programmes in Hong Kong seem to have lagged behind. A quick survey of the curricula of all translation programmes showed that The Chinese University of Hong Kong and City University of Hong Kong were the only two institutions where a technology component has been incorporated into the existing curricula. For instance, the Department of Translation at The Chinese University of Hong Kong offers courses such as "Machine Translation" and "Computers and Translation" at both the postgraduate and undergraduate levels. To better prepare our students for the new century, information technology must be incorporated in the translator-training curricula. In fact, people looking for translation jobs today will ignore information technology at their own peril. Translators stand to gain more than most other professionals from close contact with the various tools and aids that technology has to offer. As pointed out by Peck and Dorricott (1994), students must feel comfortable with the tools of the Information Age. In addition, "individuals need to learn at higher rates of effectiveness and efficiency than ever before because of rapidly growing bodies of relevant information and the escalation of knowledge and skill requirements for most jobs." (Alavi, 1994)

How should information technology be taught in an undergraduate translator-training programme in Hong Kong? Considering the local context of translator training, what should be the priorities of such programmes?

Although many academics are enthusiastic about technology and understand how it can be used creatively in the humanities, others often feel threatened by the changes in information transfer methods implicit in its use. As with any new technology, people will react against it, often in the interests of unconscious or perceived self-preservation. Students who enrol in these translator-training programmes often do so because they, too, are not attracted to technology.

So, the first problem that has to be overcome in teaching IT in translation programmes is to teach the teachers how to work with the new technology in order to help them see the advantages of incorporating IT into translator training. Teaching the student does not seem so difficult at first. However, most students will need guidance on how to make the best use of this new tool, and we should not presume that it will be as easy as some people make out.

What should be taught then? First of all, since machine translation, especially English-Chinese machine translation, still has much to improve upon, undergraduate translation programmes may be better off focusing on the utilisation of available technologies instead of teaching students rule writing. The ability to both use and interact with the Internet is particularly important for the translator, because this profession will increasingly rely on information technology not only for the process of producing translation, but for finding and exchanging information, acquiring work, and self-advertisement. Students should also be introduced to Internet resources, such as listserves, where one can request or discuss special terminology as well as ethical and professional issues. They also need to become proficient users of the World Wide Web, with its vast resources for interpreters and translators to not only gather information, but to publish their own research as well.

In addition, there are translation aids, such as electronic text corpora, translation memories, translators' workbenches, terminology databases, spell checkers, and grammar checkers. Properly understood and utilised, these tools can be of considerable help to a translator. And though many in the translation community cheerfully insist they can "pick it up as they go along," it is probably true that most people, learners and practitioners alike, would appreciate more formal training in the use of these new forms of technology. As for rule writing for translation software, we should probably reserve this for postgraduate students, who are supposed to be much more knowledgeable about translation, linguistics, and computer science.

One thing that needs to be pointed out is that making room in a translator-training curriculum for more instruction in information tech-nology will obviously alter the programme. However, this does not mean that such instruction should replace other equally important subjects. It is just that a case also needs to be made for the need to introduce technology instruction.

To sum up, a meaningful and implicative test of the comprehensibility of machine-translated texts from English to Chinese is not currently viable, despite the reported success of machine translation software between other language pairs. Bear in mind that English-Chinese translation software still has much catching up to do, compared with translation software between, say, European languages. We should not get carried away by the progress that has been achieved in machine translation as a whole, but need to seriously consider including the technological component in translator-training curricula to effectively prepare our students for today's translation market. Helping teachers and students take the first step and selecting appropriate contents for instruction seem crucial in accomplishing this goal.

To prepare translators and interpreters for the twenty-first century and to increase their competitive edge, a translating and interpreting curriculum needs to include models and principles of the basic concepts of translation and interpretation (Gile, 1995; Larson, 1987, 1991), ethics and professionalism (Hammond, 1992; Samuelsson-Brown, 1993; Gonzalez, Roseann, *et al.*, 1992; Mikkelson, 1996), translation and interpretation techniques and strategies (Baker, 1992; Dollerup and Lindegaard, 1994; Picken, 1989), and last but not the least, technological skills needed for the final product. (O'Hagan, 1996)

With all this said, it is hoped that a comprehension test of machine-translated texts from English to Chinese, which cannot be conducted with meaningful findings today, will be possible in the not too distant future.

References

Alavi, Maryam (1994). "Computer-mediated Collaborative Learning: An Empirical Evaluation." *MIS Quarterly*, Vol. 18, pp. 159–74.

Anobile, Michael (2000). Foreword, in Robert C. Sprung, ed., *Translating into Success: Cutting-edge Strategies for Going Multilingual in a Global Age*. Amsterdam and Philadelphia: John Benjamins Publishing Company.

Baker, Mona (1992). *In Other Words*. New York: Routledge.

Chan, Sin-wai (2000). "The Making of *TransRecipe*: A Translational Approach to the Machine Translation of Chinese Cookbooks." Paper presented at the International Conference on Translation and Information Technology at The Chinese University of Hong Kong, 8 December, 2000.

Clinton, Bill (2000). "A Reply to Ms. Ann G. Macfarlane." *ATA Chronicle*, 25 September.

Dollerup, Cay and Annette Lindegaard, eds. (1994). *Teaching Translation and Interpreting 2*. Amsterdam and Philadelphia: John Benjamins Publishing Company.

Fan, Y. (2000). "Me and Translation Software." *Chinese Translators' Journal*, No. 3, p. 79.

Farwell, David, Laurie Gerber and Eduard Hovy (1998). Foreword, in David Farwell, Laurie Gerber and Eduard Hovy, eds., *Machine Translation and the Information Soup*. New York: Springer-Verlag.

Gile, Daniel (1995). *Basic Concepts and Models for Interpreter and Translator Training*. Amsterdam and Philadelphia: John Benjamins Publishing Company.

Gonzalez, Roseann, Victoria Vasquez, and Holly Mikkelson (1992*). Fundamentals of Court Interpretation: Theory, Policy, and Practice*. N.C., Carolina: Academic Press.

Hammond, Deanna L., ed. (1992). *Professional Issues for Translators and Interpreters*. Amsterdam and Philadelphia: John Benjamins Publishing Company.

Larson, Mildred (1987). *Meaning-based Translation*. New York: Rodale Books.

Larson, Mildred, ed. (1991). *Translation, Theory, and Practice: Tension and Interdependence*. New York: State University of New York at Binghamton.

Leffa, Vilson J. (1994). "Machine-translated Text: Is it Comprehensible to Proficient Readers?" *System*, Vol. 22, No. 3, pp. 391–99.

Liu, Qun and Yu Shiwen (1998). "*TransEasy*: A Chinese-English Machine Translation System Based on a Hybrid Approach," in Farwell David, Gerber Laurie and Eduard Hovy, eds., *Machine Translation and the Information Soup*. New York: Springer-Verlag, pp. 514–17.

MacFarlane, Ann G. (2000). "Letter to Mr. Bill Clinton." *ATA Chronicle*, 9 March.

Maia, Belinda (1998). "LETRAC — Language Engineering for Translation Curricula." http://www.hit.uib.no/AcoHum/abs/Maia1.htm.

Neubert, Albrecht (1991). "Computer-aided Translation: Where Are the Problems?" *Target*, Vol. 3, No. 1, pp. 55–64.

O'Hagan, Minako (1996). *The Coming Industry of Teletranslation*. Clevedon: Multilingual Matters Ltd.

Peck, Kyle L. and D. Dorricott (1994). "Why Use Technology?" *Educational Leadership*, Vol. 51.

Picken, Catriona, ed. (1989). *The Translator's Handbook*. London: Aslib.

Samuelsson-Brown, Geoffrey (1993). *A Practical Guide for Translators*. Clevedon: Multilingual Matters Ltd.

Sprung, Robert C. (2000). "Introduction," in Robert C. Sprung, ed., *Translating into Success: Cutting-edge Strategies for Going Multilingual in a Global Age*. Amsterdam and Philadelphia: John Benjamins Publishing Company.

Wood, P. (2000). "What Do Translators Do? And What Machines Can Not." *The Linguist*, Vol. 39, No. 5, pp. 133–35.

Globalisation on Language: Death of the Translator in the Technological Age

Paris Lau Chi-chuen
General Education Centre
Hong Kong Polytechnic University

In his article "The Work of Art in the Age of Mechanical Reproduction," Walter Benjamin laments upon the withering of an aura in the work of art in the age of mechanical reproduction. (Benjamin, 1992:211–44) Imitation as a mode of reproduction is in fact indispensable for art. Through it pupils could practice their craft, masters could diffuse their works, and third parties could be in pursuit of gain. Yet mechanical reproduction in the modern age no longer contributes to art in the same way. Instead it functions negatively by detaching the object of art from its original space and time, from its original domain of tradition. The authority of the object is jeopardized by its replica and the unique existence of art is undermined by the plurality of copies. Transitoriness and reproducibility replace authenticity and permanence, resulting in the liquidation of the traditional value of the cultural heritage.

Benjamin's argument touches upon the intricate relationship between the original and its copies. The original is art because it is embedded in the fabric of culture and therefore asserts a cult value both unique in space and time. Copies produced mechanically lose the aura of tradition. New exhibition values could be superimposed upon them, depending on the particular situation of the beholders and spectators. Politics replaces aesthetics. Mechanical reproduction changes the reaction of the masses towards art. In the process of mass consumption, independence and individuality of the viewers are lost. They become receptive machines. The unconventional is uncritically enjoyed while the truly new is criticised with aversion. For Benjamin, the new mode of collective participation leads to certain social significance accountable to contemporary crisis and renewal of mankind.

II

The translated work is by nature an imitation of the original. The search for equivalence and correspondence between the original and its replica, or between the source language/culture and the target language/culture has been going on in the entire history of translation. Eugene A. Nida has specified two basic orientations in translating: one towards formal equivalence and the other towards dynamic equivalence. (Nida, 2000:126–40) By formal equivalence, he refers to the attempts of a translator to reproduce as literally and meaningfully as possible both the form and content of the original. By dynamic equivalence, he refers to the attempts to keep a substantial correspondence in the relationship between the response of the receptor in the source language/culture towards the original and that of the receptor in the target language/culture towards the translated work. The former principle aims at linguistic and conceptual equivalence closest to the source while the latter aims for psychological and hermeneutical correspondence in the reading process.

As an imitation of the original, the replica is bound to be inferior. Both principles of equivalence specified by Nida will simply render the translated work dependent and parasitic. Even at its best the translated work can only be a near equivalent of the original, at whatever levels of correspondence. The pledge of faithfulness already hides the tension for betrayal. Yet the ideological assumption of the translated work, albeit in another language, as an inferior copy of the original, is in fact a relatively recent phenomenon of the last century. As argued by Bassnett and Trivedi, it arose as a result of the invention of printing, the spread of literacy, and the emergence of the notion of an author as owner of the text. Such a belief coincides with the period of early colonial expansion, when Europe began to reach outside its own boundaries for territory to appropriate. (Bassnett and Trivedi, 1999:1–18)

With reference to the history of translation in modern China, the rapid development in Chinese translation of Western works in the late nineteenth century marks the same period of colonial invasion onto the Mainland proper. Translation activities did flourish in much earlier periods: from the late Han to the ninth century in the translation of Buddhist scriptures, then in the late Ming dynasty in the works of Jesuit priests, followed by the translation of Western learning in the late Qing dynasty. As Eva Hung has pointed out, from the second century till 1895, the authority of the translated text mainly depended upon the translators' strength of memory and educational/religious background. Most of the translators were foreigners who were second language users of the Chinese language. In some cases, bilingual or cross-cultural training was not even considered an important factor. The Chinese translation tradition seldom really emphasised the authority of the original as it seems to be. The authority was built up on the men — not the text. (Hung, 2000:15–37)

One should be careful and sceptical enough to avoid the misconception that translation is regarded as a science in the West, but an art in China. In fact the development of translation theory in the last century shows an opposite shift to the other side of the pendulum. Over the years, the Western world is rapidly decolonising its earlier assumptions on the relationship between the original and the replica. In the recent decade, cultural critics like Edward W. Said, Gayatri Chakravorty Spivak, Kwame Anthony Appiah and Homi Bhaba have all tried in one way or another to open the third space of in-betweenness in translation. (Carbonell, 1996:79–98) Meanwhile, the Chinese notion of translation is becoming more and more mimetic in orientation and technical in application.

The modern Chinese concept of translation seems to begin with Yan Fu's 嚴復 three difficulties of translation as proposed in his preface to the translation of Huxley's *Evolution and Ethics* in 1898. Here the cardinal principles of faithfulness (*xin* 信), intelligibility (*da* 達) and elegance (*ya* 雅) were first spelt out on general terms. Thereafter an entire century of theoretical discussion on translation in the Chinese context becomes a footnote to Yan Fu. As Zhang Jinghao has claimed in a recent conference, so far nobody could ever propose some translation criteria with more comprehensive, in-depth, well-layered, and concise synoposis than Yan Fu's. (Zhang, 2000:393–99) Faithfulness implies an orientation towards the original, intelligibility implies one towards the readers while elegance towards the style of the translated text. An ideal work of translation has to be authentic, readable and élitist. Of the relationship between the three criteria, Wong Wang Chi has argued that they are of one totality, centering upon faithfulness to the original. (Wang, 1999:87) It is quite a disappointment that so much incisive, detailed and careful reading on the part of

Wong comes to such a misleading conclusion, especially when he himself has reminded us that Yan Fu was not really accurate or loyal in his own translation of Huxley. The pledge of faithfulness already hides the tension for betrayal. If we do consider Yan Fu in the proper context as outlined by Wong, both faithfulness and elegance should be subservient to the intelligibility of meaning. The phrase "the aim of intelligibility is faithfulness" (為達即所以為信也) is only a rhetorical apology for the hidden guilt of betrayal.

Over the years, in the Chinese side of the world, the ideological formation of Yan Fu's triple focus was subtly replaced by the supreme concern for technical and empirical fidelity. In a recent research paper, Wang Jiquan 王繼權 has pointed out that in the last century, fictional translation has developed from free translation 意譯 (with abridgements or supplements) to word-for-word translation 直譯 (faithful to the original). Faithfulness or near faithfulness to the original becomes the common practice. For him, when both content and style of the original are preserved in translation, the quality will be obviously improved. It also proves that translators are now more serious about their work. (Wang, 2000:49–50) Does seriousness come with more faithfulness in translation? Another scholar Guo Yanli 郭延禮 also argues that translated literary works in modern China from 1870–1919 develops from casual adaptation through paraphrasing to literal translation and reaches maturation in the plain style of the post-May Fourth period. (Guo, 2000:83–84) Does literal translation (more faithfulness towards the original) bring forth more mature translation?

In the recent decade, in the theoretical field, Wang Ning has tried to re-formulate the three principles of Yan Fu and proposes his four ideal criteria of translation:

(1) have a close reading and thorough understanding of the original text by reading between the lines (the deep structure of the linguistic discourse) and even behind the lines (the implied cultural code and literary convention and other necessary background knowledge) and by identifying himself with the author or text;

(2) have a perfect rendition not only of fluency but also of preserving the original style without the translator's own subjective construction;

(3) polish the version until it reaches the high level of modern literary discourse of the target language without changing the original style;

(4) finally make the translated version retain the original's cultural code beyond the mere structural and linguistic level. (Wang, 1996: 47–48)

Even a cursory glance of the key phrases will show the reiterated emphases on fidelity and faithfulness: "have a close reading and thorough understanding of the original text," "identifying himself with the author or text," "preserving the original style without the translator's own subjective construction," "without changing the original style," and "retain the original's cultural code." According to Wang, the translator has to understand thoroughly the original text, preserve the style and the cultural code of the original text and identify himself with the original author. Imitation as a mode of reproduction is in fact indispensable for translation.

III

Argued elsewhere, Walter Benjamin has proposed the notion of translatability between the original in the source language and the translated work in the target language. The translator has to go back to the original to figure out the laws governing the translation. A translation which intends to perform a transmitting function cannot transmit anything but information. Yet, one which aims to serve the reader by reproducing the unfathomable, the mysterious, the poetic elements of the original only results in an inaccurate transmission of inessential content. They are two examples of bad or inferior translation. Benjamin claims that a translated work is an afterlife of the original which undergoes a maturing process of transformation and renewal. (Benjamin, 2000:11–23)

In our technological age, however, machine translation has altered the organic relationship as what mechanical reproduction had done to artistic creation in Benjamin's time. Automated computer programmes on a globalised network begin to produce near replicas of the original on the web. The artistic value is disregarded while the exhibition value is emphasised. All embedded fabrics of culture and tradition are lost, displaced by the cybernetic display of sheer information extracted and reproduced by the machine. On the web, the original space and time are altered disproportionately. The translated work reaches the infinite in space in a split second of time. In front of the screen of the computer monitor, readers enjoy the evanescent passage of translated texts, resulting in a new mode of mass consumption and collective participation.

The significance of machine translation can be more far-reaching than its immediate impact. It has subtly rewritten the cardinal principles for translation. In our technological age, partial faithfulness and overall intelligibility have priority over all other concerns. The translation process involves the substitution of one linguistic code to another and the displacement of one cultural system with another. Yet, so long as information could be grasped and understood in a glance, faithfulness to the original or intelligibility of all details is usually taken for granted. Stylistic and artistic considerations are equally remote on the web. Dialogues of daily communication through e-mail and *icq* (an internet tool that informs you who's online at any time for instant communication) are never elegant or even readable. Translation has to be quick, cheap and immediately accessible.

"What does it matter who is speaking?" — the attitude of indifference on the part of the readers towards the author's transcendental anonymity has prompted Michel Foucault to ponder over the disappearance, or death of the writer. (Foucault, 1979:141–60)

Nowadays, maybe we have to think about the death of the translator seriously in face of the automated revolutions in computer intelligence. What does it matter who is translating? Translator, as an imitator, still had a role to play in the past. In our technological age, however, in order to compete with machines, human translators have to make themselves machines. After all, what difference does it make who is translating?

Maybe we should learn from Foucault by locating "the space left empty by the author's [translator's] disappearance, follow the distribution of gaps and breaches, and watch for the openings that this disappearance uncovers." (Foucault, 1979:145) But in the end, is there a translator-function corresponding to Foucault's author-function in discourse?

References

Bassnett, Susan and Harish Trivedi (1999). "Introduction: Of Colonies, Cannibals and Vernaculars," in Susan Bassnett and Harish Trivedi, eds., *Post-colonial Translation: Theory and Practice*. London and New York: Routledge, pp. 1–18.

Benjamin, Walter (2000). "The Task of the Translator: An Introduction to the Translation of Baudelaire's Tableaux Parisiens," in Lawrence Venuti, ed., *The Translation Studies Reader*. London and New York: Routledge, pp. 11–23.

Benjamin, Walter (1992). "The Work of Art in the Age of Mechanical Reproduction," in Hannah Arendt, ed., Harry Zohn, trans., *Illuminations*. London: Fontana Press, pp. 211–44.

Carbonell, Ovidio (1996). "The Exotic Space of Cultural Translation," in Roman Alveraz and M. Carmen-Africa Vidal, eds., *Translation, Power, Subversion*. Clevedon: Multilingual Matters Ltd., pp. 79–98.

Foucault, Michel (1979). "What Is an Author?" in Josué V. Harari, ed., *Textual Strategies: Perspectives in Post-structuralist Criticism*. Ithaca, New York: Cornell University Press, pp. 141–60.

Guo, Yanli 郭延禮 (2000). 〈中國近代翻譯文學史的分期及其主要特點〉 (Periodization and the Major Characteristics of the History of Literary Translation in Modern China), in Wong Wang Chi 王宏志, ed.,《翻譯與創作：中國近代翻譯小說論》(*Translation and Creation: On Early Modern Chinese Translation of Foreign Fiction*). Beijing: Peking University Press, pp. 56–87.

Hung, Eva 孔慧怡 (2000). 〈中國翻譯傳統的幾個特徵〉(Some Characteristics of the Chinese Translation Tradition), in Eva Hung and Yang Chengshu 楊承淑, eds.,《亞洲翻譯傳統與現代動向》(*Translation in Asia: Past and Present*). Beijing: Peking University Press, pp. 15–37.

Nida, Eugene A. (2000). "Principles of Correspondence," in Lawrence Venuti, ed., *The Translation Studies Reader*. London and New York: Routledge, pp. 126–40.

Wang, Jiquan 王繼權 (2000). 〈略論近代的翻譯小說〉(Brief Comments on the Translated Fictions in the Modern Period), in Wong Wang Chi 王宏志, ed.,《翻譯與創作：中國近代翻譯小說論》(*Translation and Creation: On Early Modern Chinese Translation of Foreign Fiction*). Beijing: Peking University Press, pp. 49–50.

Wang, Ning (1996). "Toward a Translation Study in the Context of Chinese-Western Comparative Culture Studies." *Perspectives: Studies in Translatology*, Vol. 4, No. 1, pp. 47–48.

Wong, Wang Chi 王宏志 (1999).《重釋信、達、雅：二十世紀中國翻譯研究》(*Re-interpreting "Faithfulness, Intelligibility and Elegance": Translation in Twentieth-century China*). Shanghai: Oriental.

Zhang, Jinghao 張經浩 (2000). 〈也談翻譯研究的幾個問題〉(On Some Issues in Translation Studies), in Liu Ching-chih 劉靖之, *et al.*, eds.,《翻譯教學研討會論文集》(*Proceedings of the Conference on Translation Teaching*). Hong Kong: Hong Kong Translation Society, pp. 393–99.

Shall We Dance, When the Smart Machines Take Over, Virtually?

Evangeline S. P. Almberg
Department of Translation
The Chinese University of Hong Kong

An anonymous Machine Translation (MT) researcher once said, self-ironically:

> 把語言所不能表達出來的，
> 表達出來了的，是文學。
> 把語言説不清楚的，
> 説不清楚了的，是科學：
> 把語言説清楚了的，
> 又説不清楚了的，
> 是機器翻譯。

A human translator turns out the following:

Literature is that which expresses what language cannot express;
science is that which clarifies what language cannot clarify;
and machine translation is that which makes unclear what language
has made clear.

There are, of course, other ways of putting it, for example:

While literature expresses what language cannot otherwise express
and while science makes clear what language cannot make clear,
machine translation befuddles anew what language has already
made clear.

This is a competent paraphrase. The translator as reader has an adequate understanding of the message in the source language text and a good grasp of its context.

The key words here are "understanding" and "grasp." It concerns not facts alone but the wit and humour *and* self-irony.

What could a smart machine understand of the same message? I have tried three different systems, all purportedly having attained the state of the art in Chinese-English machine translation. Their speed is indeed impressive, but the results are far from producing any usable end-products.

Software One says:

Language place cannot express to come out, express, is literature.
It is not clear to say language, it is clear to say, is science.
Say language understand, it is not clear to say again, is machine translation.

Software Two:

Language can't express to come out of, / Language can't express and come out of,
Express and come out of, is a literature.
Can not remember clearly the *chu* the language of,
Say clarity of, is a science;
Say the language clarity of,
Again to caned not remember clearly the *chu* of,
Is a machine translation.

And *Software Three*:

Can't express the language out,
Have expressed out, is literature.
Say and know at the pairs of language,
Say clearly, is the science;
Language make clear,
Say clearly
Is machine translation.

The original text aside, what have the smart machines communicated to us through this jumble of words? Or is it a matter of communication at all? What could we understand from the above? By virtue of our fuzzy human perception, we might perhaps be able to make some educated guesses that it has to do with differentiating between literature, science and machine translation in relation to language. But any message more precise than this is lost on the way.

Let alone the serious deficiencies in sentence parsing and word recognition, e.g. *Software One*'s tendency to treat so-called "empty" words that indicate relations or moods as if they had stayed in their substantive nature to refer to some action, event or thing,[1] thus confusing the semotagtic word classes, *Software Two*'s over-facile resort to transference and *Software Three*'s ignorance of the need for a subject in the English despite its absence in the Chinese. It makes no attempt at representing the ambiguity or resolving the problem by disambiguation. There is also an odd predilection for articles, definite or indefinite. But we shall not go into the machines' idiosyncrasies. My point of displaying these translations done by machine is not so much to find fault (of which there is plenty) as to let them bear out, most ironically, the very substance of the original message.

Indeed, we are all aware today that MT has come to stay — if not for poetry (not yet any way for an unfathomably long time to come), then at least for domain-specific and repetitive or standardized materials such as weather reports, say, from English into French or vice versa. The sheer speed of the smart machine would be helpful enough. It can spit out, I heard, 6,000 Chinese characters per second to provide the gist of the original English text, without involving any human translator. *But*, as we have just seen, the prospect for the machine to translate even a brief statement of only 50 characters from Chinese into English is still very dim, despite the human translator's willingness to edit heavily afterwards.

Let us now look once again at the above attempts by three of the presumably brightest Chinese-English translation systems. They cannot even be considered word-for-word translations nor readable texts. And here is the crux that actually allows English-Chinese software to have the obvious edge over their Chinese-English counterparts.[2]

By virtue of its morphology and writing convention, the Chinese written language does not bear any demarcation of words as formally and semantically independent units more than the characters alone with their component radicals, and the machine has therefore very little clue to semantic units at the micro or word level. Instead, it often interprets each character as an independent unit rather than a part of another unit in a smaller context, and thus pulverizes the overall meaning into a loose mess with hardly any intelligible semantic filament to reel forth. The absence of formal segmentation between words in written Chinese thus creates a tremendous stumbling block for the machine. To overcome this is virtually a matter for artificial intelligence or simply machine intelligence.

How intelligent is a smart machine? How does it work as a translator? How the machine works depends (so far anyway, thank heavens) on how it has been programmed and what it has been programmed to do. Thus we come to the programmer.

But who is the programmer? What goes into the making of an MT programmer? We do not even know how a human translator works, despite various claims of expertise. We simply do not know how people decide to use the words they do to express the thoughts they develop. How then could we know the making of an MT programmer?

At this juncture, I feel obliged to admit that I am enjoying the privilege of being intellectual, namely, to have the curiosity to think and talk about what we do not know but nevertheless to ask the right questions and, occasionally, to suggest possible answers.

Having said that, I venture now to analyse the machine's output by working backwards, so to speak, to figure out how it might have worked. What is missing? How could it be improved? Or is there some radical change necessary of the modus operandi? In short, *how can a human translator help the machine to translate?* Not the other way round.

For MT is not only a matter of computer science nor merely that of computational linguistics. These are only necessary but not adequate factors. They have to do with MT in general, including English-Chinese MT. Specific to Chinese-English machine translation, extra attention to the contrastive analysis of Chinese and English is peremptory. In general, much remains to be done in natural language understanding.

To make MT truly viable, the input of human translators is indispensable. For what the machine lacks is, among other things, that fuzzy perception of a human being, which involves educated guesses and comprehension made possible by an enormous so-called "redundancy" of background knowledge, sometimes called intuition. True, MT may be able to crunch a huge corpus seemingly at the speed of light and the machine may also have been fed with gargantuan corpora of words, but this extension of the translation memory paradigm is of an essentially different quality to that of human memory and association.

A machine is incapable of emotions, humour, and irony and it lacks the "scenes and frames" (to borrow the words of Charles Fillmore) that make up human experiences — linguistic, cultural and otherwise. A good human translator has not only a broad and deep command of the language pair involved but also a vast and deep reservoir of knowledge and experience, that help things fall into place at the macro as well as the micro level.

To be fair to the machine, it is also true that an untalented *and* unlearned human beginner may turn out shoddy products not even comparable with MT outputs. For humans and machines might share certain deficiencies, including rigidity and ignorance. But while even machines are learning or being made to learn at great speed, humans, meanwhile, on top of trying to keep their own place, have much to teach the machines, thus creating new tasks for ourselves rather than losing ground to them. It is the subtle fluidity and flexibility required to manoeuvre among the "scenes and frames" of cultural and linguistic factors in translation that give humans the upper hand, and before this is gone, it would take what it takes for computer scientists to create and build into the machines such intelligence, based on the information and ideas provided by linguists, computational, neural and contrastive, *and* by human translators who know the art of crossing borders, until the day when machines have become virtually so educated and articulate that we shall dance at long last.

And just for the fun of it, I would like to conclude this paper by inviting the audience to read, at your leisure, another sample of the smart machine's output alongside that of a human being's. (See Appendix) It is poetry, and there, the hegemony of the machine is very clearly curbed. The smart machine can't go it alone. It takes two to tango. Shall we dance?

Notes

[1] Many Chinese *xu ci*, "empty" or function words, have originated from *shi ci*, "solid" or substantive words referring to actions/events, objects, relations or abstractions.

[2] A back translation of the English translation into Chinese recaptures much of the message and becomes almost readable and possibly editable.

References

Almberg, Evangeline. S. P. 吳兆朋 (1997). "O Brave New World, That Has Such People In't!" *Journal of Translation Studies*, No. 1, pp. 140–44.

Arnold, Doug J., Lorna Balkan, R. Lee Humphreys, Siety Meijer and Louisa Sadler (1994). *Machine Translation*. Oxford: Blackwell.

Fu, Aiping 傅愛平 (2000). 〈中國機器翻譯的概況〉 (Survey of Machine Translation in China). *Bulletin of the Department of Translation, The Chinese University of Hong Kong,* No. 6, January, pp. 6–7.

Gutknecht, Christoph and Lutz J. Rölle (1996). *Translating by Factors.* Albany, New York: State University of New York Press.

Nida, Eugene A. (1975). *Language Structure and Translation: Essays.* Selected and introduced by Anwar S. Dil. Stanford, California: Stanford University Press.

Nida, Eugene. A. and Charles Taber (1975). "Semantic Structures," in Eugene A. Nida, *Language Structure and Translation: Essays.* Selected and introduced By Anwar S. Dil. Stanford, California: Stanford University Press, pp. 102–30.

Ogden, Charles. K. and Ivor A. Richards (1949). *The Meaning of Meaning: A Study of the Influence of Language Upon Thought and of the Science of Symbolism.* London: Routledge.

O'Hagan, Minako (1996). *The Coming Industry of Teletranslation.* Clevedon: Multilingual Matters Ltd.

Snell-Hornby, Mary (1988). *Translation Studies: An Integrated Approach.* Amsterdam and Philadelphia: John Benjamins Publishing Company.

Steiner, George (1975). *After Babel: Aspects of Language and Translation.* London. New York: Oxford University Press.

Zhan, Weidong 詹衛東 (2000). 〈漢英機器系統研究與開發漫談〉(Discussion of Research and Development in Chinese-English Machine Translation Systems). *Bulletin of the Department of Translation, The Chinese University of Hong Kong,* No. 8, June, p. 7.

Appendix

〈聲聲慢〉   *Adagio*

尋尋覓覓	Seeking, searching,
冷冷清清	Freezing and forlorn,
淒淒慘慘戚戚	I sob and sigh of sorrow.
乍暖還寒時候	Days of sudden warmth and lingering cold
最難將息	Are not meant for repose.
三杯兩盞淡酒	Two or three cups of wine
怎敵他晚來風急	Are no match for the sharp gusts of night.
雁過也	Wild geese homing
正傷心	Break my heart —
卻是舊時相識	Pat, old acquaintance and all that.
滿地黃花堆積	Yellow chrysanthemums scatter in droves,
憔悴損	Worn and withered,
如今有誰堪摘	That no one now would pick.
守著窗兒	Fast by my window
獨自怎生得黑	I endure long lone hours till dark.
梧桐更兼細雨	Rain falls softly through *wutong* trees
到黃昏點點滴滴	Dripping and dripping well into dusk.
這次第	Things are such —
怎一個愁字了得	"Sad," what a helpless word!

(Original by Li Qingzhao)
(Human translation by Evangeline S. P. Almberg)

Machine Translation

Each is slow

Seek,
Desolate,
Wretched relative relative.
Barely warm but still cool time,
The most difficult to rest.
One is light for wine with the three and a two cup,
Wind is worried that how to fight him to come late?
Wild goos pass too,
Sad,

But know in old times.
The Manchus ground virgin pile up,
Wan and sallow decrease,
Who mays plucking?
Keep window,
How so raw as to be black alone it is!
The Chinese parasol tree hold currently drizzle,
To a bit at dusk.
Here order,
What one to worry about words!

(*Huajian Instant Translato*r, Chinese-English 5.0)

Sound sound is sought slowly,
Cheerless,
Miserable
First warm still cold time,
Do not most rest.
3 cups of two weak wines,
Zen resist he come to wind in night to worry?
Wild goose pass also, straight sad, is the acquaintance of old times.
Fully chrysanthemum accumulation,
Haggard harm,
Now have who can pick? Guard work window son, alone Zen give birth to
   dark?
Chinese parasol more simultaneously drizzle, to dusk a bit.
This time is, Zen one worry word get?

(*Transtar* 1.0)

Look for to look for the _V ¡M (Look for and look for the _V ¡M)
Cold and dreary,
The _Y _â miserably to miserably troubled at heart.
Suddenly a ·x to return the cold time, (Suddenly a ·x and return the cold to
me,)
The hardest will ®§ .
Three cups of two diluted wines,
«ç _Ä he come to breeze late nasty?
Will goose over also,
Is feeling sad,

How is old hour acquaint with.
Everywhere Huang spend and pile up,
Distressed ·l ¡M
The now have whothe _ô take off?
Stay by the side the window son,
Alone how get black!
Parasol tree more __ light drizzle,
Arrive evening order and order every drop.
This one after another,
A ·T word of «ç get!

(*Alexander Chinese/English Translation System* 2.6)

Notes on Contributors

Beverly Adab

Beverly Adab lectures in French and translation studies at the University of Aston, Birmingham, the United Kingdom. She teaches various undergraduate and postgraduate courses in translation theory, professional translation, terminology and LSP translation. Dr Adab has also recently been involved, within an EU Tempus programme, in training academics in four key Romanian universities and advising on the development of university translation programmes. Her publications include two books of annotated translations, intended to assist the trainee translator in developing a systematic approach to the process of translation; also a range of papers and chapters on various aspects of translation theories and their practical application. Her most recent publication is *Developing Translation Competence* jointly edited with Dr. Christina Schäffner. She is currently working on a book dealing with cross-cultural and other problems of translation in the advertising text.

Evangeline S. P. Almberg

Evangeline S. P. Almberg 吳兆朋 received her B.A. from the University of Hong Kong and her Ph.D. from the University of Stockholm. She has taught at The University of Hong Kong, the University of Stockholm, Lingnan College and The Chinese University of Hong Kong. During her sojourn in Sweden she also freelanced as a translator and interpreter for international bodies and public institutions. At present she is Professor at the Department of Translation, The Chinese University of Hong Kong.

Chan Sin-wai

Chan Sin-wai 陳善偉 teaches translation at The Chinese University of Hong Kong, where he is currently Professor and Chairman of the Department of Translation. His academic interests lie in machine translation, translation studies and bilingual lexicography.

Chang Baobao

Chang Baobao 常寶寶 obtained the degree of B.Sc. in 1992 and M.Sc. in 1995 at the Department of Computer Science, Shanxi University. His Ph.D. studies were carried out at the Department of Computer Science and Technology, Peking University in 1999, and he is now is a lecturer in the same department. His major research interests include computational linguistics, corpus linguistics and machine translation.

Carrie Chau Kam Hung

Carrie Chau Kam Hung 周錦紅 is a lecturer in Translation and Interpretation, Division of Language Studies, City University of Hong Kong. She obtained a Higher Diploma in Translation and Interpretation from Hong Kong Polytechnic and an M.A. in Translation from The Chinese University of Hong Kong. She has extensive experience in practical translation, publication and education. Her research interests include translation skills and development of multimedia packages for educational purposes.

Aman Chiu

Aman Chui 趙嘉文 was the divisional manager of Pearson Education China Ltd. (formerly known as Longman) for nine years, in charge of the dictionary and translation division, which had offices in both Hong Kong and Shanghai. He managed the editorial, marketing, and design staff in both cities. Currently he works independently as a publisher and writer. His current projects include a series of dictionaries for Hong Kong and China. With an M.A. in Language Studies from Hong Kong Baptist University, one of his research interests is terminology, in particular the management of Chinese IT terms.

Hong Qingyang

Hong Qingyang 洪青陽 is a graduate student at the Computer Science Department, Xiamen University, His academic interests lie in artificial intelligence, especially natural language processing and machine translation.

Hu Qinan

Hu Qinan 胡欽譜 is an M.Phil. student in the Department of Chinese, Translation and Linguistics, City University of Hong Kong. Before commencing her postgraduate studies she was a quality control engineer in the L & H Representative Office in Beijing, and a software engineer at the Institute of Software, Chinese Academy of Sciences. She did her undergraduate training at the Department of Computer Science and Engineering, Beijing Information Technology Institute.

Irene Ip Kwok Chun

Irene Ip Kwok Chun 葉幗珍 is a lecturer in English, Division of Language Studies, City University of Hong Kong. She received her university education in Australia and Hong Kong, specializing in English language and literature. She has extensive teaching experience with tertiary students from different disciplines. Her research interests are curriculum design, reading skills as well as the development of multimedia packages for educational purposes.

Bjorn Jernudd

Björn Jernudd 顏諾 is Chair Professor of Linguistics in the Department of English Language and Literature at Hong Kong Baptist University. His major research interests are language management in discourse and in the private sector, and language planning. He maintains a website <http://arts.hkbu.edu.hk/~bhjernudd/> which is a directory of language management organisations. A recent article is "Language Education Policy — Asia" in Bernard Spolsky, ed., *Concise Encyclopedia of Educational Linguistics*, Elsevier 1999. He is the editor of two issues of the *Journal of Asian Pacific Communication* on "Language Management and Language Problems" in 2000 (Part I, Vol. 10, No. 2) and 2001 (Part II, Vol. 11, No. 1).

Kang Shiyong

Kang Shiyong 亢世勇, who received an M.A. from Yan Tai Normal College, is an assistant professor in the Department of Chinese Language and Literature at that college. His research interests lie in the areas of Chinese syntax and computational linguistics.

Kit Chunyu

Kit Chunyu 揭春雨 received his Ph.D. in Computer Science from the University of Sheffield in 2000. He joined City University of Hong Kong as a lecturer in the Department of Chinese, Translation and Linguistics in 1996. Prior to this he received his B.Eng. in Computer Science and Technology from Tsinghua University in 1985, M.A. in Applied

Linguistics from the Chinese Academy of Social Sciences in 1988, M.Phil. in Linguistics from City University of Hong Kong in 1993 and M.Sc. in Computational Linguistics from Carnegie Mellon University in 1994. His research interests include machine learning of natural language and machine translation. He has a number of journal articles and conference papers on tokenization (esp. Chinese word segmentation) for natural language processing and unsupervised lexical learning.

Paris Lau Chi-chuen
Paris Lau Chi-chuen 劉自荃 obtained his Ph.D. from the School of Oriental and African Studies, University of London. Formerly Head and senior lecturer at the Department of English at Hong Kong Shue Yan College, he is now teaching at the General Education Centre of Hong Kong Polytechnic University. He has translated into Chinese for Taiwanese readers Christopher Norris's *Deconstruction: Theory and Practice* (1995), Lina Hutcheon's *The Politics of Postmodernism* (1996) and Bill Ashcroft *et al*'s *The Empire Writes Back* (1998).

Li Defeng
Li Defeng 李德鳳 is an assistant professor in the Department of Translation, The Chinese University of Hong Kong. He has taught translation and English in mainland China, Hong Kong and Canada. His academic interests include translation studies, translation teaching and second language learning and teaching. His present study in translation focuses on translation pedagogy and curriculum and material development in translation. He has written for leading international journals such as *Target, Meta, Babel, TESOL Quarterly* and *Teaching and Teacher Education.*

Li Shaozi
Li Shaozi 李紹滋 teaches computer science and technology at Xiamen University, China, where he is currently an associate professor of the department. His academic interests lie in artificial intelligence, natural language processing, and several other related fields of computer applications.

Li Tangqiu
Li Tangqiu 李堂秋 teaches computer science and technology at Xiamen University, China, where he is currently Chairman and Professor of the

Department of Computer Science. His academic interests lie in artificial intelligence, especially in the areas of natural language processing and machine translation.

Liu Qun

Liu Qun 劉群 is an associate professor of Computing Technology at the Chinese Academy of Sciences. He received his B.S. degree in Computer Science from the University of Science and Technology of China in 1989, and his M.S. degree in Computer Science from the Chinese Academy of Sciences in 1992. He has worked for the Academy of Sciences since 1992, and began studying for his Ph.D. at Peking University in 1999. His research interests revolve around machine translation and natural language processing.

Pan Haihua

Pan Haihua 潘海華 received his Ph.D. in Linguistics from the University of Texas at Austin in 1995. He joined City University of Hong Kong as an assistant professor at the Department of Chinese, Translation and Linguistics in July 1995. He has published two books and more than 20 journal articles on various topics. His research interests include syntactic theory, semantics, and computational linguistics. He has worked on a variety of research topics such as argument structure, aspect, focus and negation, locative inversion, the passive construction, quantification, reflexive binding, corpus-based linguistic knowledge acquisition and noun phrase extraction, machine translation, pronoun resolution, subject identification, word error detection and correction.

Sin King Kui

Sin King Kui 冼景炬 received his B.A. and M.A. from The Chinese University of Hong Kong and his Ph.D. from Southern Illinois University. He is now an associate professor at the Department of Chinese, Translation and Linguistics, City University of Hong Kong, teaching legal translation, translation theory and interpretation. He served on the Bilingual Laws Advisory Committee from 1990 to 1997 and was appointed an MBE (Member of the Most Excellent Order of the British Empire) by the British Government for his contribution to the translation of Hong Kong laws into Chinese. His research interests and publications are in the areas of language and law, and the philosophy of language.

Charlotte To
Charlotte To 杜潔儀 is a web editor. She is in charge of web contents editing, research and project coordination for an e-business solutions provider. She has also edited an entertainment portal and a B2C (business-to-consumer) website. She has an M.A. in Language Studies from Hong Kong Baptist University, and before joining the private sector was a research assistant in the Department of English at City University of Hong Kong. Her major research interest is terminology, in particular the management of terms in individuals' discourse in the field of information technology.

Jonathan J. Webster
Jonathan J. Webster is currently Associate Dean in the Faculty of Humanities and Social Sciences, and Acting Head of the Department of English at City University of Hong Kong. He received his Ph.D. in Sociolinguistics from the State University of New York at Buffalo in 1980, and prior to joining City University of Hong Kong in 1990 held positions at the National University of Singapore, and Hong Kong Baptist University. His research interests include text linguistics, computational lexicography and the application of metadata for linguistic exploration.

Yang Xiaofeng
Yang Xiaofeng 楊曉峰 is a graduate student at the Department of Computer Science, Xiamen University. His academic interests lie in artificial intelligence, especially natural language processing and machine translation.

Yu Shiwen
Yu Shiwen 俞士汶 is Professor and Director of the Institute of Computational Linguistics, Peking University. His research interests lie in the areas of computational linguistics, machine translation and information extraction.

Zhang Huarui
Zhang Huarui 張華瑞, who holds an M.S. degree, is a lecturer in the Institute of Computational Linguistics, Peking University. His research interests lie in the areas of Chinese lexical semantics and computational linguistics.

Index